Ocean Wave Energy Conversion

Michael E. McCormick
Corbin A. McNeill Professor of Naval Engineering
U.S. Naval Academy
Annapolis, Maryland

Dover Publications, Inc.
Mineola, New York

Copyright

Bibliographical Note

This Dover edition, first published in 2007, is an unabridged slightly corrected republication of the work originally published in 1981 by John Wiley & Sons, Inc., New York. The author has provided a new Preface to the Dover Edition.

Library of Congress Cataloging-in-Publication Data

McCormick, Michael E., 1936–
Ocean wave energy conversion / Michael E. McCormick. — Dover ed.
p. cm.
Originally published: New York : Wiley, c1981.
Includes bibliographical references and index.
ISBN-13: 978-0-486-46245-5
ISBN-10: 0-486-46245-5
1. Ocean wave power. 2. Ocean energy resources. I. Title.

TK1081.M39 2007
621.31'2134—dc22

2007012798

Manufactured in the United States of America
Dover Publications, Inc., 31 East 2nd Street, Mineola, N.Y. 11501

To my parents for their love and guidance during my formative years and to Dr. Robert Cohen for giving me the opportunity to pursue my interests in ocean energy conversion.

Foreword

The topic of wave energy conversion came to public attention in a paper by Steven Salter published in *Nature* in 1974. At about the same time Sir Christopher Cockerell's experiments on a wave contouring raft were reported in a paper in *New Scientist*. That this renaissance of interest in wave energy should take place in the United Kingdom is particularly appropriate because these islands are bordered on their western coasts by some of the world's roughest seas. For the same reason, in 1975 the United Kingdom's Department of Energy started an extensive program of research into various methods of extracting energy from sea waves. This program is still continuing, and a total of some 15 million pounds sterling (30 million U.S. dollars) has already been committed to the work.

As the development of ocean wave energy has continued, other countries have started their own programs. In Japan, also an island surrounded by rough seas, there has long been interest in the topic. As long ago as the 1940s, Commander Masuda invented and built a wave powered air turbine that actually generated a small amount of electricity from the waves in the Sea of Japan. The Norwegians Budal and Falnes also developed a wave energy conversion buoy and expect to have a prototype in operation by 1982/1983. Work on a number of different systems has also been done in Sweden, France, and the United States.

The first result of all this work was an indication that wave-generated electricity is likely to be expensive. The UK program faced a real crisis of confidence in 1978 when the first reference designs of the four original systems indicated power costs of 20–50 p/kWh (40-100 cents/kWh). It has been the experience of most researchers in the field that their first designs give power costs in this same range.

The wave energy community does not give up easily, however. Within a year costs were down to range 5–15 p/kWh (10–30 cents/kWh), and a wide variety of different solutions had produced this encouraging result. Recent work is showing that there is now a real possibility that, with engineering well within the scope of present-day technology, costs will even get close to the low end of this range.

I personally believe that the cost of electricity from ocean waves will approach the cost of fossil fuel energy. This will be achieved, I think, largely as a result of the creative and dedicated work of the closely integrated group of people who form the world's wave energy community.

The prize will not be easily won, however. The enormous power of the sea as it is seen in a storm must not cause us to forget that as an energy source it is really very diffuse. To capture such diffuse energy from such a hostile environment demands the highest level of design thinking and a full understanding of the hydrodynamic and engineering problems.

The contribution that Professor McCormick has made during the past few years to our understanding of these topics has indeed been enormous. It is, therefore, with the greatest pleasure that I commend to those who seek to understand this fascinating subject, this compilation of all the wisdom Mike has made available to us over the years.

CLIVE O. J. GROVE-PALMER

Program Manager
UK Wave Energy Technology Support Unit

Preface to the Dover Edition

Since the publication of the first edition of this book, published by Wiley-Interscience, in 1981, there have been many advances in the area of ocean wave energy conversion. Many of these have been the result of the European Union's sponsorship of research, development and demonstration projects in the member countries. The reader can learn of the European Union sponsored efforts, along with those in Japan, India and other countries by simply typing in "ocean wave energy conversion" in one of the search engines on the WEB.

The basics of ocean wave energy conversion do not change. This book is designed to present those basics. The discussions are presented in six chapters and four appendices. The discussions range from the basics of wave mechanics to the possible environmental consequences of wave energy conversion. The appendices include examples of wave-energy conversion patents. Throughout the book, the mathematical level is such that a reader with a good comprehension of high school mathematics and physics can easily understand the material.

The author wishes to express his appreciation to Professor Rameswar Bhattacharyya and to the late Professor Patricio A. A. Laura for recommending Dover Publications, and to Mr. John Grafton of Dover for his help and advice on this project.

The author also wishes to thank Professor Mark Harper for providing an opportunity to, again, teach the subject at the U. S. Naval Academy. Finally, a special thanks to Mr. Robert C. Murtha and Dr. Brian T. Cunningham of Ocean Energy Systems for their support of the author's research and development efforts in wave energy conversion.

Michael E. McCormick

Annapolis, Maryland
March, 2007

Preface

The purpose of this book is to present the basic concepts of ocean wave energy conversion and supporting materials in a way that can be useful to scientists, engineers, and inventors. Although new patents are granted each month on wave energy conversion ideas throughout the world, there are only nine or so basic wave energy conversion techniques. These techniques are fully described, and their uses and performances are illustrated. For those inventors with only a modicum of education in mathematics, the worked examples (with few exceptions) are presented in a manner requiring only a knowledge of algebra. For the scientist and engineer, more detailed discussions leading to the examples are presented. A categorized bibliography is presented for those interested in performing background research.

The book evolved from my various research, program management, and teaching experiences. My early research in wave energy conversion, supported by the U.S. Coast Guard, was in the area of wave-powered navigation aids. At the Energy Research and Development Administration (now the U.S. Department of Energy) Dr. Robert Cohen gave me the opportunity to formulate and direct the programs in the energy conversion of waves, currents, and salinity gradients. The information gained in these experiences was then used in the design of a senior-level course at the U.S. Naval Academy called "Ocean Energy Conversion."

I wish to express my sincere appreciation to Ms. Sharon Vaughn for her help in the preparation of the manuscript and to Mr. Eric Midboe of Gibbs and Cox, Inc. for his constructive criticism.

MICHAEL E. MCCORMICK

Annapolis, Maryland
January 1981

Contents

NOTATION xiii

1 INTRODUCTION 1

2 WAVE PROPERTIES 7

2.1 The Linear Wave, 7

2.2 The Nonlinear Wave, 15

A Stokes' Wave, 16

B The Solitary Wave, 19

2.3 Random Seas, 22

2.4 Summary, 26

3 WAVE MODIFICATION 28

3.1 Refraction, 28

3.2 Reflection, 32

3.3 Diffraction, 36

4 WAVE ENERGY CONVERSION 45

4.1 Basic Wave Energy Conversion Techniques, 45

A Heaving and Pitching Bodies, 46

B Cavity Resonators, 61

C Pressure Devices, 71

D Surging-Wave Energy Convertors, 78

E Particle Motion Convertors, 84

4.2 Advanced Techniques, 90

A Salter's Nodding Duck, 90

B Cockerell's Rafts, 101

C Russell's Rectifier, 110

D Wave Focusing Techniques, 117

4.3 Summary, 133

5 ENERGY CONVERSION, TRANSMISSION, AND STORAGE 137

5.1 Basic Electromechanical Energy Conversion Techniques, 137

A Mechanically Excited Generators, 138

B Pneumatically and Hydraulically Excited Generators, 143

5.2 Advanced Electromechanical Energy Conversion Techniques, 149

A Linear Inductance, 149

B Piezoelectricity, 153

C Protonic Conduction, 158

5.3 Power Scaling, 160

5.4 Energy Transmission and Storage, 163

A Electrical Cables, 164

B Energy-Intensive Products, 166

5.5 Summary, 167

6 ENVIRONMENTAL AND MOORING CONSIDERATIONS 170

6.1 Environmental Considerations, 170

A Open Ocean Operation, 170

B Coastal Zone Operation, 175

6.2 Mooring and Anchoring, 180

A Deep Ocean Operation, 181

B Coastal Zone Operation, 190

APPENDIXES 196

A Bibliography, 196

B Some Wave Energy Conversion Patents, 202

C Glossary, 226

D Conversion Factors, 230

INDEX 231

Notation

Common to All Chapters

A_{wp}	waterplane area (ft^2 or m^2)
b	crest width (ft or m)
c	phase velocity or celerity (ft/sec or m/sec)
c_g	group velocity (ft/sec or m/sec)
C	hydrostatic restoring moment (ft-lb-sec/rad or m-N-sec/rad)
d	body draft or depth to top of submerged body (ft or m)
$e^{(\ )}$	$2.178^{(\ )}$, the exponential function
E	wave energy (lb-ft or N-m)
f	wave frequency ($1/T$) (Hz)
f_z	heaving natural frequency (Hz)
f_θ	pitching natural frequency (Hz)
g	gravitational constant (32.2 ft/sec^2 or 9.81 m/sec^2)
h	water depth (ft or m)
H	wave height (ft or m)
H_s	significant wave height (ft or m)
$I_{w'}$	added-mass moment of inertia (ft-lb-sec^2/rad or m-N-sec^2/rad)
$I_{y'}$	mass moment of inertia (ft-lb-sec^2/rad or m-N-sec^2/rad)
j	$\sqrt{-1}$
k	wave number $(2\pi/\lambda)$(1/ft or 1/m)
L	body length (ft or m)
m	body mass (lb-sec^2/ft or kg)
$m_{w'}$	added mass (lb-sec^2/ft or kg)
p	wave pressure (lb/ft^2 or N/m^2)
P	wave power (ft-lb/sec or W or kW)
r	radial coordinate (ft or m)
S_T	wave period spectral density (ft^2/sec or m^2/sec)
SWL	still water level

t	time (sec)
T	wave period (sec)
T_s	significant wave period (sec)
T_z	heaving natural period (sec)
T_θ	pitching natural period (sec)
u, w	horizontal and vertical water particle velocity components (ft/sec or m/sec)
U	wind velocity measured at $z = 10$ m (ft/sec or m/sec)
V	wind velocity measured at $z = 19.5$ m (ft/sec or m/sec)
x, y, z	coordinates with origin on SWL, z is upward (ft or m)
β	angle between wavefront and bottom contour (deg)
δ	phase angle between wave-induced moment components (deg)
Δ_z	dimensionless damping factor in heaving
Δ_θ	dimensionless damping vector in pitching
ϵ	efficiency
η	free-surface displacement (ft or m)
θ	pitching angular displacement (deg)
λ	wave length (ft or m)
π	3.142
ρ	mass density of saltwater (2.00 lb-sec^2/ft^4 or 1030 kg/m^3)
σ_z	phase angle between free-surface displacement and heaving displacement (deg)
σ_θ	phase angle between free-surface displacement and pitching displacement (deg)
ω	circular wave frequency ($2\pi f$) (rad/sec)
ω_z	circular heaving frequency ($2\pi f_z$) (rad/sec)
ω_θ	circular pitching frequency ($2\pi f$) (rad/sec)

Subscripts

0	deep water conditions
b	breaking condition
i	number of a single wave

Superscripts

—	spatial average
∧	temporal average
•	time differentiation, d/dt

Chapter 2 Notation

A	parameter in equation (2.40)
B	parameter in equation (2.41)
E_k	wave kinetic energy (lb-ft or N-m)
E_p	wave potential energy (lb-ft or N-m)

Chapter 3 Notation

α	angle between wave orthogonal and barrier (deg)
K_D	diffraction coefficient of equation (3.9)
K_R	refraction coefficient of equation (3.4)
K_S	shoaling coefficient of equation (3.3)

Chapter 4 Notation

Section 4.1, A

B_1	width of a heaving body (ft or m)
$B_{(x)}$	width of a body section (ft or m)
D	diameter of a heaving circular cylinder (ft or m)
E_{KZ}	kinetic energy of a heaving body (lb-ft or N-m)
E_{PZ}	potential energy of a heaving body (lb-ft or N-m)
E_Z	total energy of a heaving body (lb-ft or N-m)
$E_{K\theta}$	kinetic energy of a pitching body (lb-ft or N-m)
$E_{P\theta}$	potential energy of a pitching body (lb-ft or N-m)
E_θ	total energy of a pitching body (lb-ft or N-m)
F_0	wave force amplitude (lb or N)
F_z	vertical wave force component (lb or N)
F_{zc}	vertical wave force on circular cylinder (lb or N)
F_{zR}	vertical wave force on rectangular cylinder (lb or N)
K_I	added-mass moment of inertia coefficient
K_m	added-mass coefficient
M_0	wave moment amplitude (lb-ft or N-m)
$M_{\theta c}$	wave moment on circular cylinder (lb-ft or N-m)
$M_{\theta R}$	wave moment on rectangular cylinder (lb-ft or N-m)
M_θ	wave-induced moment (lb-ft or N-m)
N	an integer
P_z	power in heaving motion (lb-ft/sec, W, or kW)

P_θ power in pitching motion (lb-ft/sec, W, or kW)
R radius of a circular float (ft or m)
Z height of a float (ft or m)

Subscripts

0 amplitude
z heaving
θ pitching

Sections 4.1, B and 4.1, C

A_1 cross-sectional area of the water column (ft^2 or m^2)
A_2 cross-sectional flow area of the turbine (ft^2 or m^2)
A_c conduit cross-sectional area (ft^2 or m^2)
D_0 diameter of float (ft or m)
D_1 diameter of the water column (ft or m)
D_2 diameter of the turbine passage (ft or m)
f_c cavity resonant frequency (Hz)
F_{PD} pressure force (lb or N)
F_{PD0} pressure force amplitude (lb or N)
H_1 double amplitude of the water column motion (ft or m)
L_1 stillwater water column length (ft or m)
p_1 pressure on the large piston (lb/ft^2 or N/m^2)
p_2 pressure on the small piston (lb/ft^2 or N/m^2)
T_c cavity resonant period ($1/f_c$)(sec)
v_1 water column velocity (ft/sec or m/sec)
v_2 air velocity in turbine passage (ft/sec or m/sec)
W_P pump work (lb-ft or N-m)
z_1 vertical displacement of the water column (ft or m)
Δ developed head (ft or m)
φ velocity potential (ft^2/sec or m^2/sec)
ω_c resonant circular frequency of the cavity ($2\pi f_c$)(rad/sec)

Sections 4.1, D and 4.1, E

A_d projected deflector area (ft^2 or m^2)
B_d width of the deflector (ft or m)
B_c width of the flap (ft or m)
E_f energy of the flap (lb-ft or N-m)
F_d force on the deflector (lb or N)
F_c force on the flap (lb or N)

F_p force transmitted to the piston (lb or N)
l_f stroke of the piston (ft or m)
P_d power of the deflector (lb-ft/sec, W, or kW)
P_p power converted by the water wheel (lb-ft/sec, W, or kW)
s_d deflector displacement (ft or m)
S_{d_0} deflector stroke (ft or m)
T_p resisting torque (lb-ft or N-m)
V_d deflector velocity (ft/sec or m/sec)
V_0 deflector velocity amplitude (ft/sec or m/sec)
$\bar{z}_f$ center of pressure (ft or m)
Γ deflector angle (deg)
ω_p rotational frequency of water wheel (rad/sec)

Section 4.2, A

f_D natural frequency of the duck $(1/T_D)$(Hz)
K_D design constant
L_D length of the duck above water (ft or m)
M mass of the duck (lb-sec^2/ft or kg)
P_D power available to the duck (lb-ft/sec, W, kW)
r_D radial position of the center of gravity (ft or m)
$R_0(z)$ paunch "radius" (ft or m)
R_1 rotor radius (ft or m)
R_2 stern radius (ft or m)
R_D beak radius (ft or m)
T_D natural period of the duck (sec)
Λ design angle of the duck (deg)

Section 4.2, B

$A_{a,b}$ motion amplitudes of hinges *a* and *b* (ft or m)
$A_{O,L}$ motion amplitudes of bow (O) and stern (L) (ft or m)
E_H energy of raft system (lb-ft or N-m)
K dimensionless wave number *kh*
$M_{1,2,3}$ wave moments on rafts 1, 2, and 3 (lb-ft or N-m)
$R_{a,b}$ reaction forces on hinges *a* and *b* (lb or N)
$z_{a,b}$ vertical displacements of hinges *a* and *b* (ft or m)
$z_{O,L}$ vertical displacements of bow (O) and stern (L) (ft or m)
$\alpha_{a,b}$ damping coefficients at hinges *a* and *b* (lb-ft-sec/rad or N-m-sec/rad)
$\xi_{1,2,3}$ vertical displacements of raft centers (ft or m)

Subscripts

$1, 2, 3$	raft numbers
a, b	hinge numbers
O, L	bow (O) and stern (L)

Section 4.2, C

A	flow area (ft^2 or m^2)
B	width of catch basin (ft or m)
B_{cell}	width of a single cell (ft or m)
E_R	potential energy of the water column (lb-ft or N-m)
L	length of rectifier (ft or m)
$\xi_{u,l}$	upper (u) and lower (l) reservoir levels (ft or m)
Δ_R	hydraulic head (ft or m)

Section 4.2, D

d_I	island (atoll) depth (ft or m)
d_0	maximum depth of atoll (ft or m)
D_1	water column diameter (ft or m)
F_0	vertical wave force amplitude (lb or N)
H'	transmitted wave height (ft or m)
l	body separation length (ft or m)
P_0	power available to heaving body (lb-ft/sec, W, or kW)
P_{opt}	optimum absorbed power (lb-ft/sec, W, or kW)
r	radial coordinate (ft or m)
r_1	float radius (ft or m)
r_0	island (atoll) radius (ft or m)
R'	radius of lee face of lens (ft or m)
R''	radius of front face of lens (ft or m)
$R_0(\omega)$	radiation resistance (lb-sec/ft or N-sec/m)
V_0	amplitude of vertical velocity (ft/sec or m/sec)
X_1	focal length with lee face curvature only (ft or m)
X_2	focal length with two-face curvature (ft or m)
α_L	turning angle of wave orthogonal (deg)
β_L	see Figure 4.44
β_0	see Figure 4.44
θ	angular coordinate (deg)
θ_0	angular coordinate at $r = r_0$ (deg)
ψ	phase angle between force and motion (deg)

Chapter 5 Notation

Section 5.1, A

e instantaneous voltage (V)
e_0 voltage amplitude (V)
$f_{1,3}$ frequency of sprockets 1 and 3 motions (Hz)
i instantaneous electrical current (A)
i_o electrical current amplitude (A)
P_o instantaneous electrical power (W or kW)
$r_{1,2,3}$ radii of sprockets 1, 2, and 3 (ft or m)
v_0 vertical cylinder velocity (ft/sec or m/sec)
$v_{1,2,3}$ linear gear velocities of sprockets 1, 2, and 3 (ft/sec or m/sec)
Z_o heaving amplitude (ft or m)
Φ phase angle between voltage and current (deg)

Section 5.1, B

A_1 capture chamber (cavity) cross-sectional area (ft^2 or m^2)
A_2 turbine flow area (ft^2 or m^2)
H_1 internal wave height (ft or m)
p_0 downstream ambient pressure (lb/ft^2 or N/m^2)
$p_{1,2}$ cavity (1) and upstream turbine (2) pressure (lb/ft^2 or N/m^2)
P available power to turbine (lb-ft/sec, W, or kW)
P_T turbine power (lb-ft/sec, W, or kW)
Q volume rate of airflow (ft^3/sec or m^3/sec)
r_i inner blade radius (hub) (ft or m)
r_0 outer blade radius (tip) (ft or m)
U_T linear midspan blade velocity (ft/sec or m/sec)
$V_{1,2}$ cavity (1) and upstream (2) turbine vertical air velocities (ft/sec or m/sec)
w_2 absolute inlet air velocity (ft/sec or m/sec)
Y_T turbine blade coefficient of equation (5.24)
α_2 blade angle (deg)
η_1 displacement of internal free surface (ft or m)
$\varphi_{1,2}$ cavity (1) and turbine (2) air velocity potentials (ft^2/sec or m^2/sec)
ψ_2 upstream turning angle (deg)
ω_T angular velocity of the turbine (rad/sec)

Section 5.2, A

B_e magnetic induction (Wb/ft^2 or Wb/m^2)
F_e motion-resisting force (lb or N)

k_e spring constant (lb/ft or N/m)
l_e wire length (ft or m)
m_e mass of the magnets (lb-sec^2/ft or kg)
N_e number of wire turns
P_e power delivered by linear-inductance device (W or kW)
P_z power of a heaving body (lb-ft/sec, W, or kW)
R_e load resistance, (Ω)
x_e displacement of the magnets (ft or m)
X_0 amplitude of magnetic motion (ft or m)
ξ relative displacement (ft or m)
φ_e phase angle between magnetic motions and buoy motions (deg)
ω_e circular frequency of magnetic motions (rad/sec)

Section 5.2, *B*

b_{pc} width of the crystal (ft or m)
d_{pc} piezoelectric strain constant (ft/V or m/V)
E_m mechanical energy (lb-ft or N-m)
E_{pc} piezoelectric energy (lb-ft or N-m)
g_{pc} piezoelectric stress constant (V-ft/lb or V-m/N)
l_{pc} length of the crystal (ft or m)
L_{pc} length of crystal stack (ft or m)
Y_{pc} modulus of elasticity of the crystal (lb/ft^2 or N/m^2)
δ_{pc} thickness of the crystal (ft or m)
Δ_{pc} stack separation (ft or m)
ϵ_{pc} dielectric permittivity (lb/V^2 or N/V^2)
Σ summation symbol

Section 5.2, *C*

D diameter of protonic conductor and piston (ft or m)
E_p energy delivered by protonic conductor (lb-ft or N-m)
F Faraday's constant (96,500 coulombs)
p_l pressure of hydrogen in lower chamber
p_u pressure of hydrogen in upper chamber (lb/ft^2 or N/m^2)
R universal gas constant (8.32 W-sec/mole-°K)
T_a absolute temperature (°K)

Section 5.3

A area (ft^2 or m^2)
C proportionality constant

f frequency $(1/T)$(Hz)
n length scale factor
P power (lb-ft/sec, W, or kW)
T period (sec)
V velocity (ft/sec or m/sec)

Subscripts

m model
p prototype

Chapter 6 Notation

Section 6.1, A

F fetch length (nautical miles)
$P_{a,b}$ wave power corresponding to conditions a and b in Figure 6.2 (lb-ft/sec, W, kW)
P_0 deep water wave power (lb-ft/sec, W, or kW)
P_1 transmitted wave power (lb-ft/sec, W, or kW)
x_B separation length of systems A and B (ft or m)
x_C separation length of systems B and C (ft or m)

Section 6.1, B

H_0' transmitted wave height (deep water) (ft or m)
P_b wave power at breaker line (lb-ft/sec, W, or kW)
P_b' conversion-affected wave power at breaker line (lb-ft/sec, W, or kW)
P_l longshore wave power in surf zone (lb-ft/sec, W, or kW)
P_l' conversion-affected longshore wave power in surf zone (lb-ft/sec, W, of kW)
α_b angle between wave front and breaker line (deg)

Section 6.2, A

D terminal diameter (ft or m)
G vertical position of catenary origin (ft or m)
k_c spring constant of mooring line (lb/ft or N/m)
$l_{1,2,3}$ lengths of cable segments 1, 2, and 3 (ft or m)
l_c length of elastic cable (ft or m)
n number of mooring lines

r_c radius of cable (ft or m)
Y_c modulus of elasticity of cable (lb/ft^2 or N/m^2)

Section 6.2, B

F_a line tension at anchor (lb or N)
F_0 line tension at float (lb or N)
K_a holding power of anchor
w_c cable weight per unit length (lb/ft or N/m)
W_a weight of anchor (lb or N)
X, Z coordinate system of catenary (ft or m)
φ_a mooring line angle at anchor (deg)
φ_0 mooring line angle at float (deg)

1 Introduction

The most conspicuous form of ocean energy is the *surface wave*. Waves are simply energy in transition, that is, the energy being carried away from its origin. The sources of wave energy are the following four phenomena: (1) bodies moving on or near the surface causing relatively low period waves of low energy; (2) winds generating seas and swells; (3) seismic disturbances causing the (misnamed) "tidal wave" or "tsunami"; (4) the lunar and solar gravitational fields causing the largest waves, the tides (the actual "tidal wave"). The relative energies of these waves have been estimated by Munk et al. (1957) and are shown in Figure 1.1. The tides are predictable, and wind-generated waves are also predictable if the nature of the wind is known. From the results in Figure 1.1 we see that these two wave types also have the highest relative wave energies. In this book our attention is focused on the conversion of *wind-wave energy* into more usable forms. Tidal energy conversion has been discussed in a number of publications, the more notable of which is the book edited by Gray and Gashus (1972) and the report by Wayne (1977).

Wind waves are actually a form of *solar energy* since the primary source of wind energy is the sun. Solar radiation is collected by both land and water masses, and the water is the more efficient collector of the two. The air above a warmed water mass is then heated. The warm air rises into the higher elevations replacing the cooler, more dense air which, in turn, descends. Thus thermal air currents are generated. In addition to these vertically oriented currents, wind circulation patterns are established in which the warm air above the equatorial waters rises and moves toward the polar regions where the air is cooled, descends, and again flows toward the equator. These wind circulation patterns are modified by both the presence of land masses and the rotation of the earth. The resulting global circulation patterns of the wind are sketched in Figure 1.2. For more thorough discussions of the meteorological aspects of wind generation, the reader is referred to the books of Voss (1972) and Dietrich (1963).

It is interesting to note that in each transformation of the solar energy the power intensity (power per unit area normal to the direction of energy transmission) increases. For example, the average solar insolation at a latitude of 15° N is 0.170 kW/m^2. The wind velocity at this latitude in the mid-Pacific

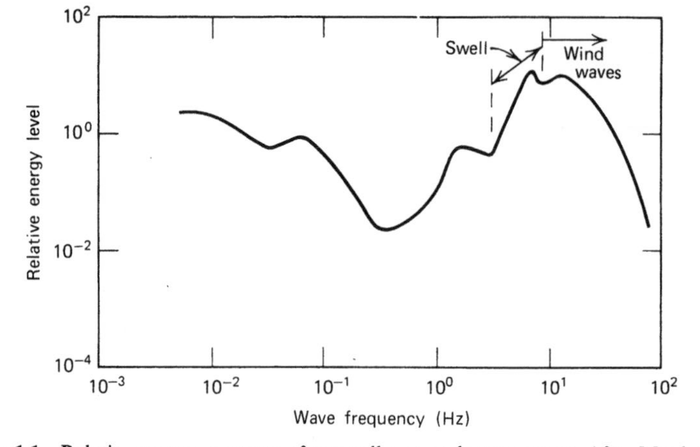

Figure 1.1 Relative energy spectrum of naturally created water waves. After Munk (1957).

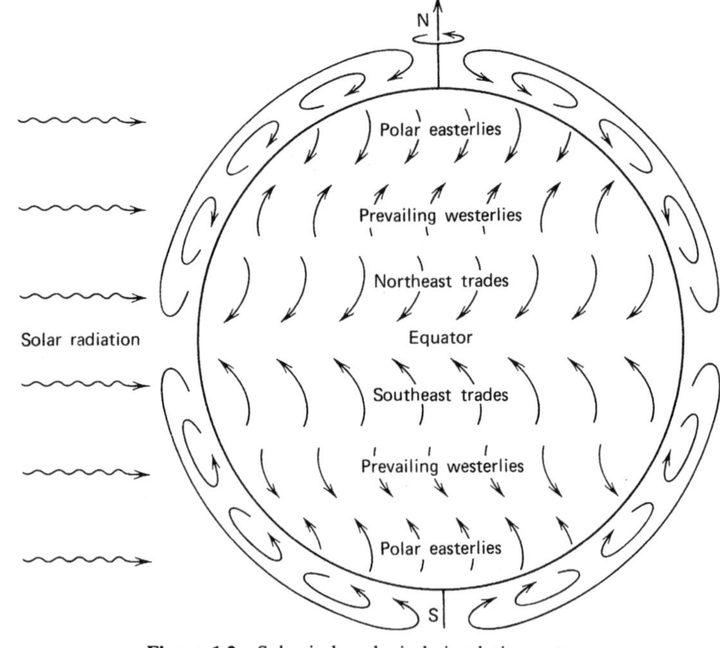

Figure 1.2 Solar-induced wind circulation patterns.

ocean may be approximately 20 knots (10 m/sec) within the Northeast Trades and thus have a power intensity of 0.580 kW/m^2. The average wave generated by this wind has a power intensity of 8.42 kW/m^2. The averaged power densities, on a global scale, however, are reversed. Estimates of the total powers of the five ocean energy forms are given by Wick and Isaacs (1976) and are shown in Figure 1.3.

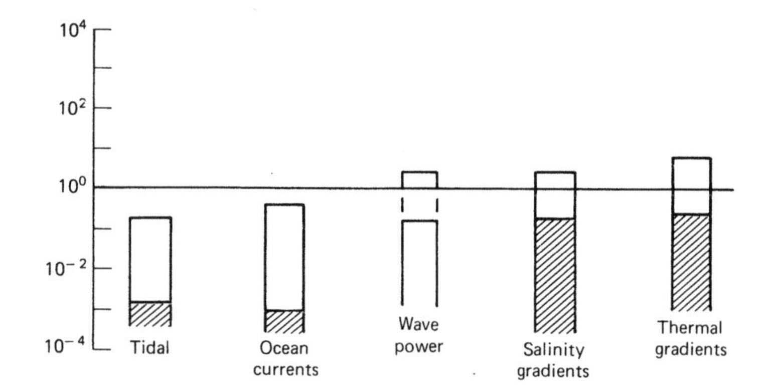

Figure 1.3 Power or energy flux for various sources of ocean energy sources. After Wick and Isaacs (1976).

Since wave energy is so conspicuous, many inventive individuals have been inspired to devise methods of converting the energy of ocean waves into more usable forms. For example, on March 1, 1898 P. Wright patented the devise shown in Figure 1.4. An operating system was also constructed by Bouchaux-Praceique in France in the early part of the twentieth century and is described by Palme (1920). A sketch of this system, taken from the paper by McCormick (1976), is shown in Figure 1.5. Referring to Figure 1.5, we see that the

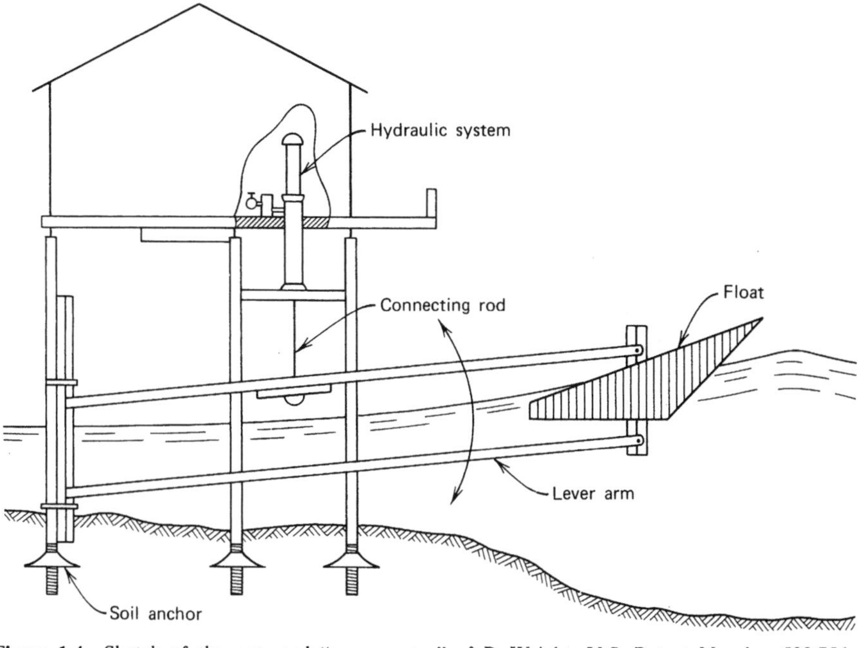

Figure 1.4 Sketch of the patented "wave motor" of P. Wright, U.S. Patent Number 599,756, March 1, 1898. The outrigger configuration.

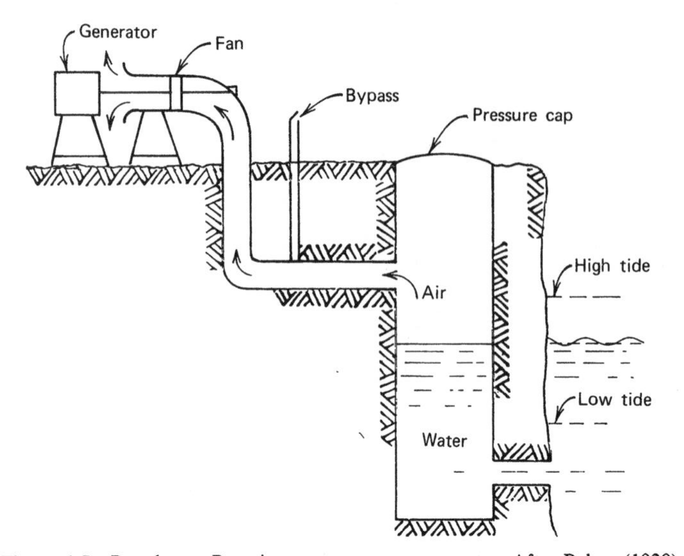

Figure 1.5 Bouchaux–Praceique wave energy convertor. After Palme (1920).

rise and fall of the water surface excites the air column above it which, in turn, drives the turbogenerator. The system was said to supply all of Bouchaux-Praceique's electrical power requirements. The schemes shown in Figures 1.4 and 1.5 are two of many wave energy conversion ideas. In fact, there are over 1000 patented wave energy conversion techniques in Japan, North America, Western Europe, and the United Kingdom. Most of these patented ideas are variations of a few very basic ideas described in Chapter 4 of this book. Appendix B contains a partial list of patent numbers referring to wave energy conversion schemes.

The *economics* of wave energy conversion depend on three considerations: (1) the magnitude and dependability of the wave resource, (2) the cost of construction and maintenance of the conversion system, and (3) the energy transmission from the site to the user. McCormick (1976) calculated the wave power per crest length along the coasts of the continental United States on the basis of observed wave height and period data presented in the U.S. Army's Shore Protection Manual (1973). The results of these calculations are presented in Figure 1.6, where the seasonal variation of the monthly averaged wave power per crest length is shown. The results in Figure 1.6 suggest that a significant resource exists only along the Northwest coastline. Pierson and Salfi (1976) show that the magnitude of the resource in deep waters well away from the Northwest coast are an order of magnitude greater than the values shown in Figure 1.6. *Dispersive and shoaling processes*, described in Chapter 3, reduce the power per crest length of waves (generated in deep water) by the time these waves reach the coast. On the basis of the resource estimates, therefore, one can conclude that wave energy conversion is more feasible in deep waters than in coastal waters. The problem arising in deep water wave

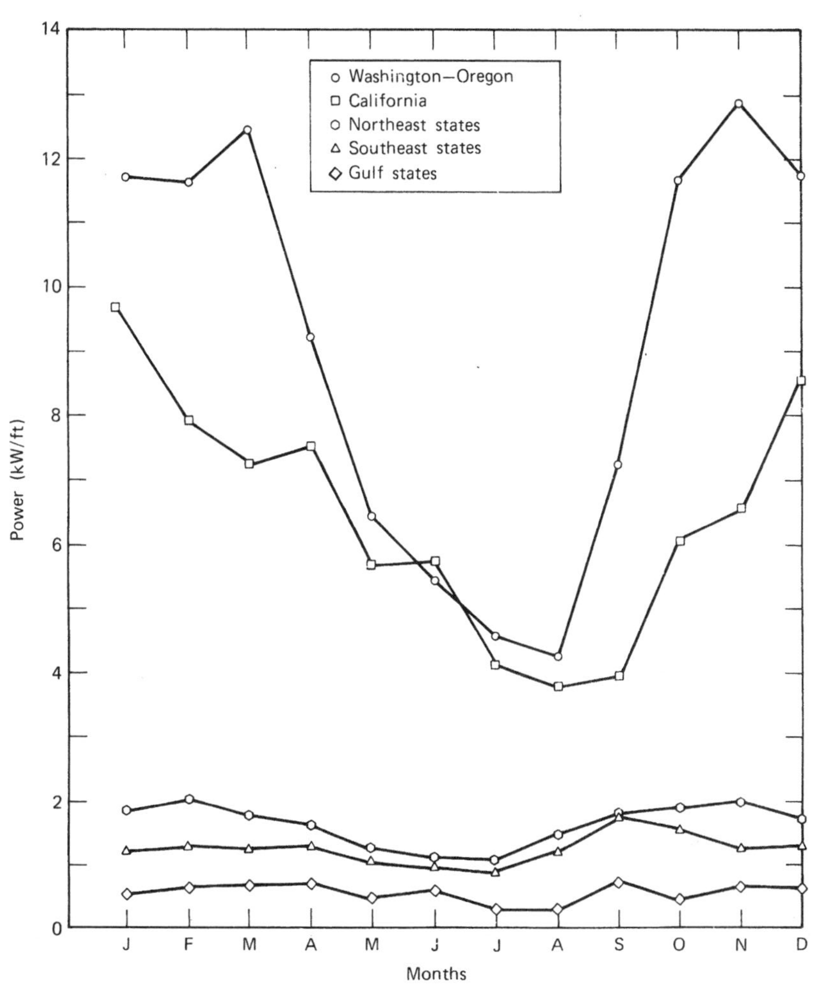

Figure 1.6 Wave power variations over 1 year. Data from coastal waters, but calculations based on deep water assumptions. After McCormick (1976).

energy conversion is in finding an economical method of energy transmission from the conversion site to the energy market. This problem also exists in ocean thermal energy conversion (OTEC), where the resource is primarily in equatorial waters, that is, between 20° N latitude and 20° S latitude. In connection with OTEC, energy transmission has been thoroughly studied, and the reader is referred to the publication by Konopka et al. (1977) for more information on this subject. Direct electrical transmission is possible up to 80 miles or 128 km. Furthermore, for energy conversion sites located more than 80 miles (128 km) from shore, the manufacturing of energy-intensive products, such as aluminum, is possible. Thus deep water wave energy conversion is not only feasible but is probably cost effective.

All of the topics mentioned in this introductory chapter are more thoroughly discussed in the chapters that follow. The reader is also encouraged to consult the listed references at the end of each chapter for more detailed discussions of the various topics.

References

Dietrich, G. (1963), *General Oceanography*, Wiley-Interscience, New York.

Gray, T. J., and Gashus, O. K., Eds. (1972), *Tidal Power*, Plenum, New York.

Konopka, A., Talib, A., Yudow, B., Blazek, C., and Biederman (1977), "Alternate Energy Transmission Systems from OTEC Plants," U.S. Department of Energy Report No. DSE/2426-20.

McCormick M. E. (1976). "Salinity Gradients, Tides and Waves as Energy Sources," Energy from the Oceans Conference, North Carolina State University, Raleigh, UNC-SG-76-04, January, pp. 33–41.

Munk, W., Tucker, M., and Snodgrass, F. (1957), "Remarks on the Ocean Wave Spectrum," National Academy of Sciences, Publication No. 515, Washington, D.C., pp. 45–60.

Palme, A. (1920), "Wave Motion Turbine," *Power*, Vol. 52, No. 18, pp. 200–201.

Pierson, W. J., and Salfi, R. E. (1976), "The Temporal and Spacial Variability of Power from Ocean Waves Along the West Coast of North America," Wave and Salinity Gradient Workship Proceedings, University of Delaware, Paper E.

U.S. Army (1973), *Shore Protection Manual*, U.S. Government Printing Office, Stock No. 0822-00077, Washington, D.C.

Voss, G. L. (1972), *Oceanography*, Golden Press, New York.

Wayne, W. W. (1977), "Tidal Power Study for the United States Energy Research and Development Administration," Stone and Webster Engineering Corporation, Boston, Mass.

Wick, G. L., and Isaacs, J. D. (1976), "Utilization of the Energy from Salinity Gradients," Wave and Salinity Gradient Energy Workshop, University of Delaware, Paper A.

2

Wave Properties

The two measurable properties of water waves are the *height* and the *period*. These properties are measured by using an assortment of *wave gauges* or, in some situations, estimated by sight. Much of the recorded wave data actually result from *visual observations*. Researchers found that the observed wave heights did not correspond to the average wave height, but more nearly to the average of the one-third highest waves. This statistically averaged wave is called the *significant wave*.

In this chapter our interest is focused on the basic mathematical description of the water wave. To ensure that the reader obtains a working knowledge of the material presented herein, a number of numerical examples are presented after each topic. It should be noted that there is no mathematical theory that exactly describes the behavior of water waves. The various wave theories simply approximate, to some degree, the actual phenomena. For the purposes of this book, we use the results of the simplest of theories known as *Airy's theory* or the *linear wave theory*. Usable results of this theory are presented without derivation. For those readers interested in the research aspects of wave mechanics, the books by Wiegel (1964) and McCormick (1973) and that edited by Ippen (1966) are recommended.

2.1 The Linear Wave

The *swell* is a wave of relatively great length and small height. When a storm rages at sea, waves of various heights and periods are generated; however, as these waves travel away from the storm region, the waves of low period disappear leaving only waves of high period (i.e., swells). The swell is the most frequently observed wave and, as such, is of primary interest in wave energy conversion. Although the *wave height* H is much smaller than the *wavelength* λ, the relative energy of the swell is high, as can be seen in Figure 1.1.

For a wave height to wavelength ratio H/λ of $1/50$ or less, the *linear theory* can be used with excellent accuracy in predicting the kinematic properties of waves. Referring to Figure 2.1, the mathematical expressions for the *free-surface displacement* and the *wave period* are, respectively (McCormick,

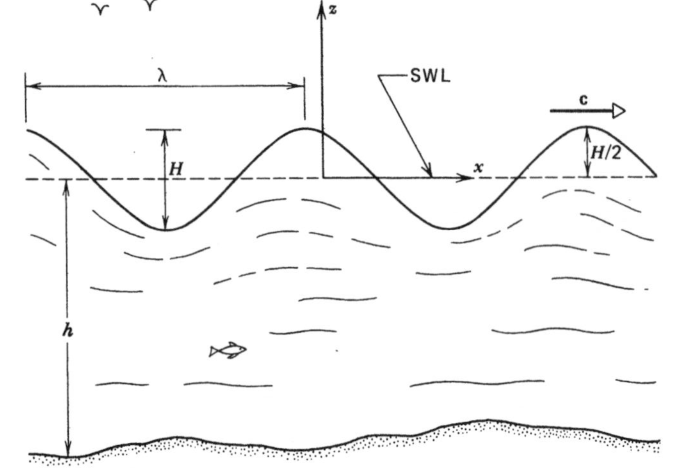

Figure 2.1 Nomenclature of a water wave. A linear wave—one having a sinusoidal profile.

1973)

$$\eta = \frac{H}{2}\cos\left(\frac{2\pi x}{\lambda} - \frac{2\pi t}{T}\right) \tag{2.1}$$

and

$$T = 2\pi\left[\frac{2\pi g}{\lambda}\tanh\left(\frac{2\pi h}{\lambda}\right)\right]^{-1/2}$$

$$= \frac{1}{f}$$

$$= \frac{2\pi}{\omega} \tag{2.2}$$

where tanh( ) is the hyperbolic tangent, f is the wave *frequency*, ω is the circular wave frequency ($2\pi f$), g is the *gravitational constant* (32.2 ft/sec^2, 9.81 m/sec^2), and h is the water *depth*. The period T is normally considered to be invariant with both *time* t and depth h; however, this is not true over long travel distances. Equation (2.2) can be rearranged to obtain the expression for the *wavelength*:

$$\lambda = \frac{gT^2}{2\pi}\tanh\left(\frac{2\pi h}{\lambda}\right) \tag{2.3}$$

In equation (2.3) λ is on both sides of the equality sign and cannot be isolated. Equation (2.3) must, therefore, be solved by using either graphic or numerical techniques. For the convenience of the reader, the solution of equation (2.3) is

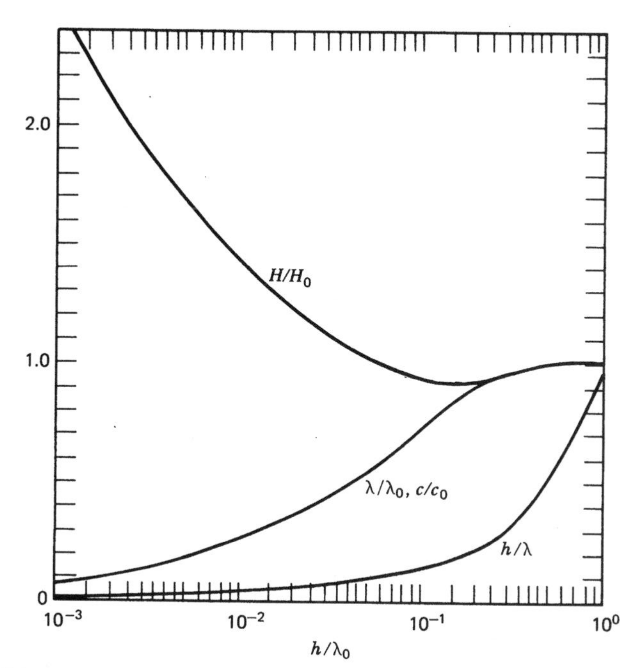

Figure 2.2 Dimensionless wave properties as functions of the depth to deep water wavelength ratio as predicted by linear theory.

presented in Figure 2.2 in dimensionless form along with the dimensionless expressions of the wave *height* H and the *phase velocity* or the *celerity* c. The properties having the subscript 0 in Figure 2.2 (i.e., λ_0, H_0, and c_0) are those occurring in *deep water*. Deep water is assumed when the depth is greater than half of the wavelength (i.e., $h > \lambda/2$). The results in Figure 2.2 indicate that this is not exactly correct; however, the assumption is satisfactory for the purpose of this book.

Individual waves travel at a *phase velocity* c. This velocity is described by the equation

$$c = \frac{\lambda}{T} = \frac{gT}{2\pi}\tanh(kh) \tag{2.4}$$

where k is the *wave number* defined by

$$k = \frac{2\pi}{\lambda} \tag{2.5}$$

At this point the reader should note that the wave height H does not appear in equations (2.2) through (2.4). The *kinematic properties* of linear waves are functions of period and depth only.

As is stated in Chapter 1, wave energy conversion appears to be feasible in *deep water*. For engineering purposes, deep water is defined where $h/\lambda \geqslant \frac{1}{2}$, or

$kh \geqslant \pi$. In this case equations (2.3) and (2.4), respectively, are approximated by

$$\lambda = \frac{gT^2}{2\pi} \tag{2.6}$$

and

$$c = \frac{gT}{2\pi} \tag{2.7}$$

Example 2.1

(a) Waves of 5-sec period are observed in deep water. From equations (2.6) and (2.7), respectively,

$$\lambda = 128 \text{ ft } (39.0 \text{ m})$$

and

$$c = 25.6 \text{ ft/sec } (7.8 \text{ m/sec})$$

(b) Ten-second waves in deep water have the following length and phase velocity:

$$\lambda = 512 \text{ ft } (156 \text{ m})$$

and

$$c = 51.2 \text{ ft/sec } (15.6 \text{ m/sec})$$

The *shallow water* approximations of the wavelength and phase velocity are, respectively,

$$\lambda = \sqrt{gh}\, T \tag{2.8}$$

and

$$c = \sqrt{gh} \tag{2.9}$$

Equations (2.8) and (2.9) are used when values of h/λ are $\frac{1}{20}$ or less.

Example 2.2

(a) When the 5-sec wave enters the *shoaling region* (where the depth affects λ and c) equations (2.8) and (2.9) can be used when the depth is 2.01 ft (0.613 m) or less.

(b) The validity range of these equations for the 10-sec wave is where the depth is 8.05 ft (2.45 m) or less.

Example 2.3

(a) The length and the phase velocity of the 5-sec wave in 2.01 ft (0.613 m) of water are, respectively,

$$\lambda = 40.2 \text{ ft } (12.3 \text{ m})$$

and

$$c = 8.05 \text{ ft/sec } (2.45 \text{ m/sec})$$

from equations (2.8) and (2.9).

(b) For the 10-sec wave in 8.05 ft (2.45 m) of water

$$\lambda = 161 \text{ ft } (49.1 \text{ m})$$

and

$$c = 16.1 \text{ ft/sec } (4.91 \text{ m/sec})$$

A comparison of the results in Examples 2.1 and 2.3 shows that the *wavelength and the phase velocity both decrease significantly as the wave approaches the shoreline.* Theoretically, both λ and c approach zero values as h approaches zero.

The water particles within the wave travel with *horizontal* and *vertical velocity components* of

$$u = \frac{\pi H}{T} \frac{\cosh[k(z+h)]}{\sinh(kh)} \cos(kx - \omega t) \tag{2.10}$$

and

$$w = \frac{\pi H}{T} \frac{\sinh[k(z+h)]}{\sinh(kh)} \sin(kx - \omega t) \tag{2.11}$$

respectively. The *deep water* approximations of equations (2.10) and (2.11) are

$$u = \frac{\pi H}{T} e^{kz} \cos(kx - \omega t) \tag{2.12}$$

and

$$w = \frac{\pi H}{T} e^{kz} \sin(kx - \omega t) \tag{2.13}$$

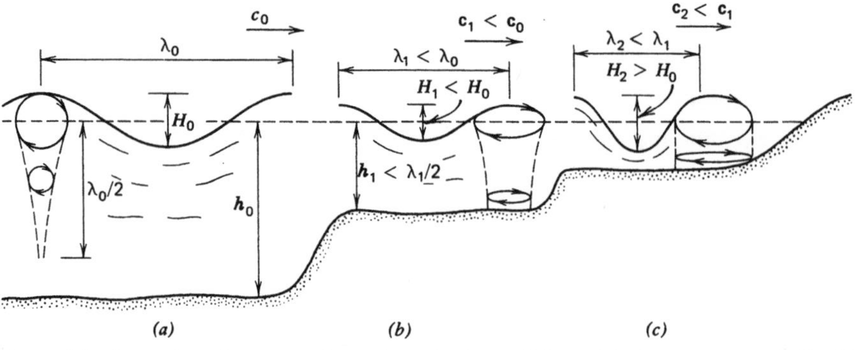

Figure 2.3 Properties of waves under various depth conditions: (*a*) deep water, $h \geqslant \lambda/2$; (*b*) intermediate water, $\lambda/2 > h \geqslant \lambda/20$; (*c*) shallow water, $h < \lambda/20$.

In *shallow water*, where $h/\lambda < \frac{1}{20}$, the approximate expressions are

$$u = \frac{H}{2}\sqrt{\frac{g}{h}}\cos(kx - \omega t) \tag{2.14}$$

and

$$w = \frac{\pi H}{T}\frac{(z+h)}{h}\sin(kx - \omega t) \tag{2.15}$$

Equations (2.10) and (2.11) describe water particle motions in waters of *intermediate depth*, where $\frac{1}{2} < h/\lambda < \frac{1}{20}$. The paths of these particles are elliptic with major and minor axes that decrease with the depth coordinate z (see Figure 2.3*b*). In *deep water* the particles travel in circular paths with diameters that decrease exponentially with z. The reader should keep in mind that the depth coordinate z is negative beneath the *still water level* (SWL). Deep water paths are shown in Figure 2.3*a*. In *shallow water* the particles travel in elliptic paths with constant major axes and with minor axes that decrease with z. Referring to Figure 2.3*c*, the particles on the free surface travel as far and as fast in the horizontal direction as do those adjacent to the sea floor.

Example 2.4

(a) The maximum horizontal and vertical particle velocity components for a wave with a height of 3 ft (0.914 m) and a period of 10 sec in deep water obtained from equations (2.12) and (2.13), respectively, are

$$u_{max} = 0.960 \text{ fps } (0.292 \text{ mps})$$

and

$$w_{max} = 0.942 \text{ fps } (0.287 \text{ mps})$$

where u_{max} occurs when $\cos(kx - \omega t) = 1$ and $z = H/2$, (i.e., at a wave *crest*) and w_{max} occurs when $\sin(kx - \omega t) = 1$ and $z = 0$ (i.e., a *node*) and fps represents feet per second and msp, meters per second.

(b) The wave that has the same height and period as that in part a where $h = 8.05$ ft (2.45 m) is just into shallow water. Here, equations (2.14) and (2.15) respectively, yield

$$u_{max} = 3.00 \text{ fps } (0.914 \text{ mps})$$

and

$$w_{max} = 0.942 \text{ fps } (0.287 \text{ mps})$$

The results in Example 2.4 indicate that the surface particle velocity actually increases as the depth decreases. Since the wave's phase velocity c has decreased with decreasing depth, at some point the maximum horizontal velocity component of the surface particles u_{max} will equal the phase velocity; that is

$$u_{max}\big|_{z=H/2} = c \tag{2.16}$$

This condition is called the *break*. As the wave travels further to shore, u_{max} exceeds c, the wave *spills*, and the wave energy is lost to turbulence and friction; hence broken waves are not of major interest in wave energy conversion.

As derived by McCormick (1973) and others, the *total energy* in a wave is obtained from

$$E = E_p + E_k = \frac{\rho g H^2 \lambda b}{8} \tag{2.17}$$

where b is the width of the crest, ρ is the mass density of water (1.93 slugs/ft^3 or 1000 kg/m^3 for freshwater and 2.00 slugs/ft^3 or 1030 kg/m^3 for saltwater). The total energy in deep water in a wave described by linear theory is equally composed of *potential energy* E_p and *kinetic energy* E_k; thus

$$E_p = E_k = \frac{\rho g H^2 \lambda b}{16} \tag{2.18}$$

The potential energy is exhibited by the wave height H, whereas the kinetic energy is dependent on the motions of the particles.

The transfer of wave energy from point to point in the direction of wave travel is characterized by the *energy flux* or, more commonly, *wave power*:

$$P = \frac{\rho g H^2 c_g b}{8} \tag{2.19}$$

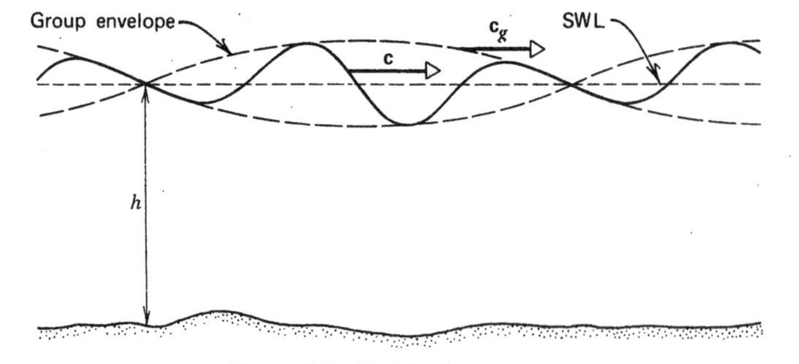

Figure 2.4 Sketch of a wave group.

where c_g is called the *group velocity* and is represented by

$$c_g = \frac{c}{2}\left\{1 + \frac{2kh}{\sinh(2kh)}\right\} = nc \qquad (2.20a)$$

When a patch of waves is observed traveling in *deep water*, that is, where $h \geqslant \lambda/2$, waves appear at the end of the patch, travel to the front of the patch, and then disappear. In the case of a deep water wave patch or group the relationship between the phase velocity and the group velocity is

$$c_g = \frac{c}{2} \qquad \text{(deep water)} \qquad (2.20b)$$

In *shallow water* ($h \leqslant \lambda/20$), however, the waves remain stationary with respect to the group boundaries; thus

$$c_g = c \qquad \text{(shallow water)} \qquad (2.20c)$$

A wave group is illustrated in Figure 2.4.

Example 2.5

(a) The total energy per unit crest width of a 3-ft (0.914-m) wave of 10-sec period in deep saltwater is

$$\frac{E}{b} = 37{,}130 \text{ ft-lb/ft } (165{,}200 \text{ J/m})$$

from equation (2.17). The power per unit crest width of this wave, from equation (2.19), is

$$\frac{P}{b} = 1856 \text{ ft-lb/sec/ft } (8.26 \text{ kW/m})$$

(b) Consider the wave in part a just entering shallow water where $h = 8.05$ ft

(2.45 m). Equation (2.17) yields a total energy per unit crest width of

$$\frac{E}{b} = 11{,}600 \text{ ft-lb/ft } (51{,}890 \text{ J/m})$$

and equation (2.19) yields the power per unit crest length:

$$\frac{P}{b} = 1166 \text{ ft-lb/sec/ft } (5.19 \text{ kW/m})$$

The expressions presented in this section can be used to obtain fair approximations of the wave properties. It must be remembered, however, that the linear theory is based on two assumptions: (1) there are no energy losses due to friction, turbulence, and so on; and (2) the wave height H is much smaller than the wavelength λ. As previously mentioned, the latter is a property of a swell that is of primary interest in wave energy conversion.

2.2 The Nonlinear Wave

One failing of the linear wave theory described in Section 2.1 is that it always predicts a *sinusoidal profile* such as that sketched in Figure 2.1. A deep water swell having a low value of H/λ will be well approximated by this profile; however, as the wave begins to shoal, that is, to be affected by the seafloor, the wave profile will begin to change to one with a narrow crest and broad trough, as sketched in Figure 2.5. This profile is said to be *nonlinear*. The reader should note that the SWL and the *mean water level* (MWL) coincide for the linear wave of Figure 2.1; however, the SWL will be below the MWL for the nonlinear wave since the MWL is defined as being half the distance from trough to crest. Since the position of the SWL is defined by the water depth h, this is the most logical level on which to place the origin of our coordinate system.

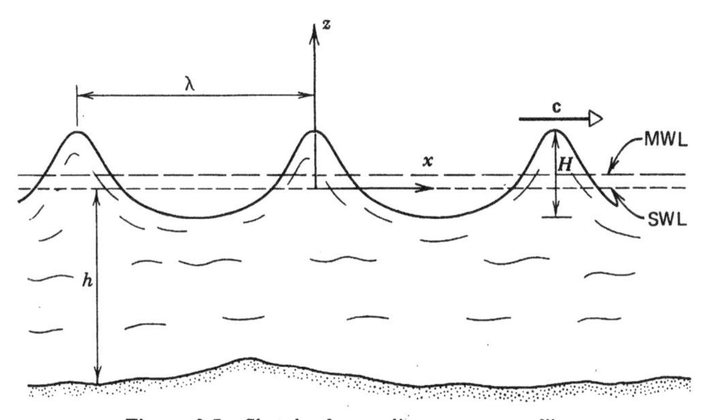

Figure 2.5 Sketch of a nonlinear wave profile.

In this section we discuss two of the nonlinear wave theories, *Stokes' second order theory* and the *solitary wave theory*. More thorough discussions of these and other nonlinear theories can be found in Wiegel (1964). Our interest in the nonlinear theory is in its ability to more accurately predict wave behavior in *shallow water*.

A Stokes' Wave

Stokes (1847, 1880) introduces an irrotational water wave theory that utilizes series representations of wave properties. The accuracy of the theory depends on the number of terms contained in the series. For example, *Stokes's first order theory* is identical with the linear theory discussed in Section 2.1. Stokes' second order theory improves the accuracy in determining the wave profile and, in addition, the *mass transport convection velocity* and the *breaking condition* of the wave. Stokes' third and higher order theories successively add to the accuracy of the wave profile prediction. For our purposes, the second order theory is satisfactory.

Without derivation, the *shallow water wave profile* predicted by Stokes' second order theory is obtained from

$$\eta = \frac{H}{2} \text{cox}(kx - \omega t) + \frac{3}{16} \frac{H^2}{k^2 h^3} \cos[2(kx - \omega t)] \tag{2.21}$$

where k is the wave number of equation (2.5) and ω is the circular frequency $2\pi f$. By comparing equations (2.1) and (2.21), the reader can see that the first term on the right-hand side of equation (2.21) is the expression for the linear profile; thus the second term is simply a *correction* to the first order (linear) theory. The expressions for the *wavelength* λ and the *phase velocity* or the *celerity* c are the same as those predicted by the linear theory, that is, those expressions found in equations (2.3) and (2.4), respectively.

The expression for the horizontal water *particle velocity* in shallow water according to the *linear theory* is

$$u = \frac{\omega H}{2kh} \cos(kx - \omega t) \tag{2.22}$$

whereas that according to Stokes' second order theory is

$$u = \frac{\omega H}{2kh} \cos(kx - \omega t) + \frac{3}{16} \frac{\omega H^2}{k^3 h^4} \cos[2(kx - \omega t)] \tag{2.23}$$

As discussed in Section 2.1, a wave is said to *break* when the particle velocity at a crest and the phase velocity are equal; that is

$$u_{\max}\big|_{z=H/2} = c \tag{2.16}$$

At a crest one can assume that

$$\cos(kx - \omega t) = \cos\left[2(kx - \omega t)\right] = 1$$

Thus if the results of equations (2.9) and (2.22) are used, the *breaking condition* in *shallow water* according to the *linear theory* is

$$\frac{\omega H_b}{2kh} = \sqrt{gh}$$

or, incorporating the results of the equation (2.8), the *breaking height* is

$$H_b = 2h \tag{2.24}$$

Combining equations (2.9), (2.16), and (2.23) results in the *breaking condition* according to Stokes' second order theory. The result is

$$\frac{\omega H_b}{2kh}\left[1 + \tfrac{3}{8}\frac{H_b}{k^2h^3}\right] = \sqrt{gh}$$

or, using the result of equation (2.8), the breaking height is

$$H_b = \frac{16\pi^2h^2}{3gT^2}\left[-1 + \sqrt{1 + \frac{3gT^2}{4\pi^2h}}\right] \tag{2.25}$$

Thus H_b of equation (2.25) depends on both the period and the depth in shallow water, whereas the breaking height of equation (2.24) is independent of wave period.

After the wave breaks the wave becomes a *surge*, which is no longer approximated by wave theories. This surging motion is of interest to those interested in the *wave energy conversion technique* described in Section 4.1, D.

Example 2.6

Consider a shallow water wave having a period T of 10 sec in 6 ft (1.83 m) of water. The breaking height obtained from *linear theory* [equation (2.24)] yields

$$H_b = 2h$$

$$= 2(6)$$

$$= 12 \text{ ft } (3.66 \text{ m})$$

The breaking height expression of Stokes's second order theory [equation

(2.25)] yields

$$H_b = \frac{16\pi^2 h^2}{3gT^2}\left[-1 + \sqrt{1 + \frac{3gT^2}{4\pi^2 h}}\right]$$

$$= \frac{16\pi^2(6)^2}{3(32.2)(10)^2}\left[-1 + \sqrt{1 + \frac{3(32.2)(10)^2}{4\pi^2 6}}\right]$$

$$= 3.22 \text{ ft } (0.981 \text{ m})$$

Obviously, the breaking wave height predicted by the linear theory is unrealistic since the trough would be on the seafloor. Furthermore, the profile of the linear breaking wave is *sinusoidal*, which is, again, unrealistic. The actual profile is sharp crested, as illustrated in Figure 2.6.

The expressions for the *total wave energy* and the *wave power* in *shallow water* obtained using Stokes' second order theory are, respectively,

$$E = \frac{\rho g H^2 \lambda b}{8}\left[1 + \frac{9}{64}\frac{H^2}{k^4 h^6}\right] \tag{2.26}$$

and

$$P = \frac{\rho g H^2 c_g b}{8}\left[1 + \frac{9}{64}\frac{H^2}{k^4 h^6}\right] \tag{2.27}$$

where b is the crest width and the group velocity c_g is equal to the phase velocity c in shallow water, from equation (2.20c). By comparing equation (2.26) with equation (2.17) and equation (2.27) with equation (2.19), it can be seen that Stokes' second order theory simply adds a *correction factor* $\frac{9}{64}H^2/k^4h^6$ to the energy and power expressions of the linear theory.

Example 2.7

The 10-sec breaking wave in Example 2.6 has a breaking height H_b of 3.22 ft (0.981 m) in 6 ft (1.83 m) of water. The shallow water wave power

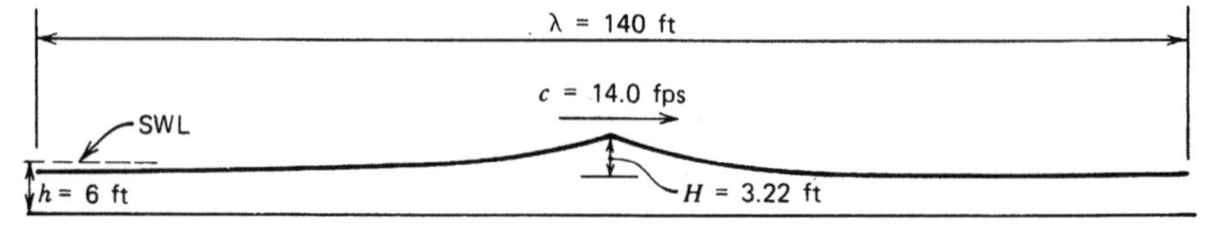

Figure 2.6 Shallow water breaking profile as predicted by the Stokes's second order theory in Example 2.6.

per crest width b according to linear theory [equation (2.19)] is

$$\frac{P}{b} = \frac{\rho g H^2 c}{8} = \frac{\rho g H^2 \sqrt{gh}}{8}$$

$$= \frac{2.00(32.2)(3.22)^2\sqrt{(32.2)6}}{8}$$

$$= 1160 \ \frac{\text{ft-lb/sec}}{\text{ft}} \quad (5.16\ \text{kW/m})$$

The Stokes' second order expression [equation (2.22)] yields

$$\frac{P}{b} = \frac{\rho g H^2 c}{8}\left[1 + \frac{9}{64}\frac{H^2}{k^4 h^6}\right]$$

$$= \frac{\rho g H^2 \sqrt{gh}}{8}\left[1 + \frac{9}{64}\frac{H^2}{k^4 h^6}\right]$$

$$= 1160\left[1 + \frac{9}{64}\frac{(3.22)^2}{(0.0452)^4(6)^6}\right]$$

$$= 1160[1.075] = 1247 \ \frac{\text{ft-lb/sec}}{\text{ft}} \quad (5.55\ \text{kW/m})$$

where the results of equations (2.5), (2.8), and (2.9) are used. The correction to the linear theory power expression is 7.5%. The expression of the correction factor in shallow water

$$\frac{9}{64}\frac{H^2}{k^4 h^6} = \frac{9}{64}\frac{H^2(gh)^2 T^4}{(2\pi)^4 h^6}$$

is a strong function of both wave height H and period T. Thus moderate increases in either of these wave properties will result in significant values of the correction to the linear theory results.

B *The Solitary Wave*

From the sketch in Figure 2.6, it can be seen that a relatively long swell in shallow water can be approximated as an isolated phenomenon. Furthermore, as the swell nears the surf zone the free surface of the wave can be entirely above the SWL. This phenomenon, called *wave set up*, is due to a "piling up" of the water particles by the convection of the shallow water waves and the

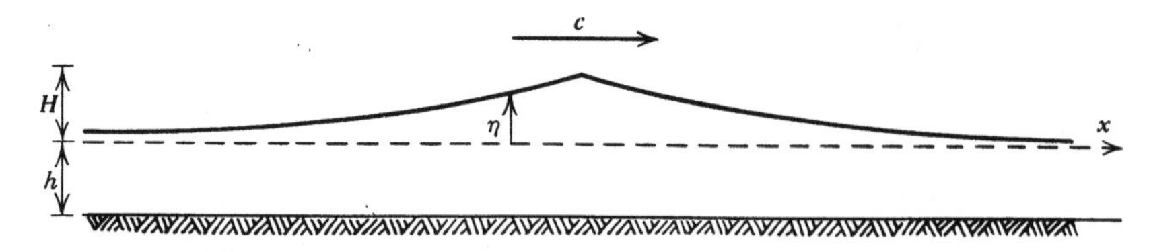

Figure 2.7 Profile of a solitary wave near a breaking condition.

surging motion of the broken waves. In this section we analyze the "isolated" swell near the surf zone by using the so-called solitary wave theory, which is thoroughly discussed elsewhere in literature (Wiegel, 1964; U.S. Army, 1973).

A sketch of a solitary wave is shown in Figure 2.7. The reader should note that the wave height H is now the vertical distance between the SWL and the wave crest. There is no wave period in this case since the wave is treated as an isolated event. Without derivation, the displacement of the free surface above the SWL is obtained from

$$\eta = H \operatorname{sech}^2\left[\left(\frac{3}{4}\frac{H}{h^3}\right)^{1/2}(x - ct)\right] \tag{2.28}$$

where c is the celerity or phase velocity. A comparison of the expressions of equations (2.1), (2.21), and (2.28) shows the basic difference between the solitary wave and the periodic wave. That is, a trigonometric term indicates periodicity, whereas the hyperbolic function indicates an isolated phenomenon.

The expression for the *phase velocity* obtained from the solitary theory is

$$c = \left[gh\left(1 + \frac{H}{h}\right)\right]^{1/2} \tag{2.29}$$

which shows that the wave slows down as the depth of the water h decreases. The *horizontal component of the water particle velocity* is

$$u = \sqrt{\frac{g}{h}}\, H \operatorname{sech}^2\left[\left(\frac{3}{4}\frac{H}{h^3}\right)^{1/2}(x - ct)\right] \tag{2.30}$$

The expression for the *breaking condition* is that derived by Laitone (1960):

$$H_b = 0.714 h_b \tag{2.31}$$

Again, after the wave breaks, the wave energy is transformed into *turbulence* and thus dissipated into heat. The region between the outermost break and the farthest extent of the uprush on the beach is called the *surf zone*. Because of the rapid energy dissipation, this region is not considered to be suitable for *wave energy conversion.*

The expression for the *pressure* beneath a solitary wave is

$$p = \rho g(\eta - z) \tag{2.32}$$

The pressure at any point within the water is simply a *hydrostatic pressure.*

Finally, the expression for the *energy* per crest width of a solitary wave is

$$\frac{E}{b} = 1.54\rho g(Hh)^{3/2} \tag{2.33}$$

Example 2.8

The deep water wave described in Example 2.5a has a height H of a 3 ft (0.914 m) and a 10-sec period T. The energy per crest width from that example is 37,130 ft-lb/ft (165,200 N-m/m). Assume very little loss in energy as this wave shoals. Thus the deep water energy value is also that of equation (2.33):

$$\frac{E}{b} = 1.54\rho g(Hh)^{3/2}$$

$$= 37{,}100 \text{ ft-lb/ft } (165{,}000 \text{ N-m/m})$$

At the breaking of the wave, equation (2.31) must be satisfied. Combining equations (2.31) and (2.33) with the preceding energy value yields

$$\frac{E_b}{b} = 1.54\rho g(0.714h_b^2)^{3/2}$$

$$= 37{,}100 \text{ ft-lb/ft } (165{,}000 \text{ N-m/m})$$

The *breaking depth* is then

$$h_b = \left[\frac{1.08}{\rho g}\left(\frac{E_b}{b}\right)\right]^{1/3}$$

$$= 8.54 \text{ ft } (2.60 \text{ m})$$

This value combined with equation (2.31) yields the *height of the breaking wave*

$$H_b = 0.714h_b$$

$$= 6.10 \text{ ft } (1.86 \text{ m})$$

2.3 Random Seas

In Chapter 1 it is stated that the direct energy source of ocean waves is the *wind*. Thus the wave height, the period, and the direction must be related to the velocity and the direction of the wind. To illustrate, a mild breeze will generate waves of small height and low period that generally travel in the direction of the breeze. Storm-generated seas, however, include waves of various heights and periods that travel in many directions.

On a windy day the waves might be observed to be *random* in height, period, and direction. Although appearing to be purely random, there are rather well defined relationships between the wave properties and the wind speed and direction. As shown in Chapter 4, most wave energy conversion devices are period dependent since they operate most efficiently at their own natural period. The efficiency of operation in a real or random sea, however, will be less than that in a *monochromatic* of a single-period sea. McCormick (1978) presents a theoretical analysis of the wave power transmitted to a linear array of ideal wave energy conversion devices. The term "ideal" means that the devices convert all the wave power into usable power; that is, the devices are 100% efficient. McCormick's analysis is an extension of that presented by Pierson and Salfi (1976). In this section the method of analysis of a random sea is both outlined and illustrated.

A *random sea* can be considered to be composed of numerous *linear waves* of various heights and periods, each traveling in a particular direction. For the ith wave, the energy obtained from equation (2.17) is

$$E_i = \tfrac{1}{8}\rho g H_i^2 \lambda_i b \tag{2.34}$$

where b is an arbitrary crest width. To evaluate the energy of all the component waves passing a point, we use Pierson's technique (1955) and represent the wave height term of equation (2.34) by

$$\frac{H_i^2}{8} = S_T(T_i)\,\delta T_i \tag{2.35}$$

where $S_T(T_i)$ is called the *wave spectral density* or, more commonly, the *wave spectrum* and δT_i is the *elemental wave period*. The units of $S_T(T_i)$ are $\mathrm{ft}^2/\mathrm{sec}$ or $\mathrm{m}^2/\mathrm{sec}$.

In most of the random wave literature, the wave spectrum is expressed in terms of frequency rather than period. In equation (2.2) the wave period is expressed in terms of frequency. Using that expression, we write

$$T_i = \frac{1}{f_i} = \frac{2\pi}{\omega_i} \tag{2.36}$$

The differential element of T_i is then

$$\delta T_i = -\frac{\delta f_i}{f_i^2} = -2\pi\frac{\delta\omega_i}{\omega_i^2} \tag{2.37}$$

The wave period spectrum of equation (2.35) can, therefore, be expressed in terms of the frequency spectrum $S_\omega(\omega)$, as follows:

$$S_T(T_i)\,\delta T_i = -\frac{2\pi}{\omega_i^2}\,S_T(T_i)\,\delta\omega_i$$

$$= -S_\omega(\omega_i)\,\delta\omega_i \qquad (2.38)$$

In this book the choice of the period spectrum is consistent with our concentration of the use of wave properties that are directly measurable—the wave height H_i and the wave period T_i.

There are a number of empirical expressions of $S_T(T)$. The two most generally used are those of Pierson and Moskowitz (1964), describing a fully developed sea, and Bretschneider (1952), describing a developing sea. We use the *Pierson–Moskowitz spectrum* for the purpose of illustration since it is based on open ocean measurements of a *fully developed sea*. Both the Pierson–Moskowitz and Bretschneider spectra have the general form of

$$S_T(T_i) = AT_i^3 e^{-BT_i^4} \qquad (2.39)$$

where for the Pierson–Moskowitz spectrum

$$A = 8.10 \times 10^{-3}\,\frac{g^2}{(2\pi)^4} \qquad (2.40)$$

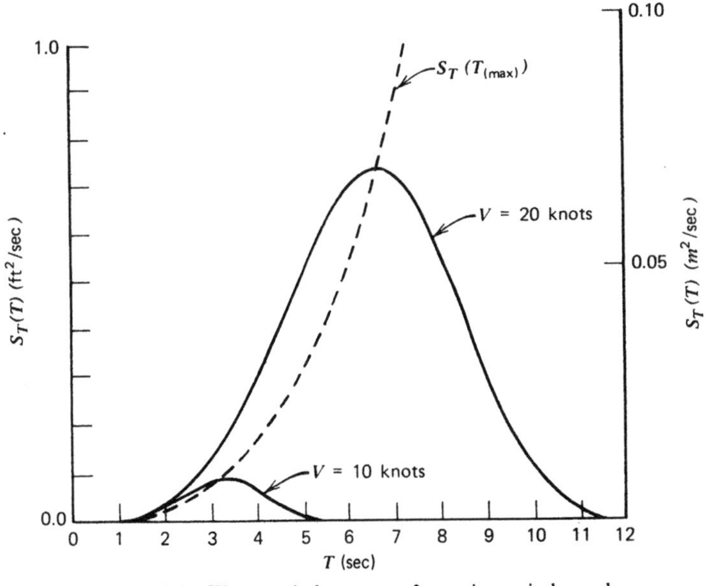

Figure 2.8 Wave period spectrum for various wind speeds.

and

$$B = 0.74\left(\frac{g}{2\pi V}\right)^4 \tag{2.41}$$

where V is the wind velocity measured 64.0 ft (19.5 m) above the SWL. Examples of the behavior of $S_T(T)$ are shown in Figure 2.8.

We now divide equation (2.34) by $\lambda_i b$; combine the resulting expression with equations (2.35), (2.39), (2.40), and (2.41); and sum (integrate) the result over all wave periods to obtain the expression of the wave *energy per unit surface area*; that is

$$\begin{aligned}\bar{E} &= \rho g \int_0^\infty S_T(T)\,dT \\ &= \rho g A \int_0^\infty T^3 e^{-BT^4}\,dT \\ &= \frac{\rho g A}{4B} \\ &= 2.74 \times 10^{-3}\,\frac{\rho V^4}{g}\end{aligned} \tag{2.42}$$

where the mass density ρ of saltwater is 2.00 lb-sec^2/ft^4 or 1030 kg/m^3 and the gravitational acceleration g is 32.2 ft/sec^2 or 9.81 m/sec^2.

Example 2.9

At a location in the Northeast Tradewind Belt (see Figure 1.2) the wind speed is approximately a steady 20 knots or 33.8 f/sec (10.3 m/sec). The sea is, then, fully developed in deep water locations well away from the land masses. The Pierson–Moskowitz spectrum [equation (2.39)] for this wind speed is shown in Figure 2.8. By equating the first derivative of equation (2.39) to zero, we obtain the value of the wave period corresponding to the *peak energy spectrum*. That is, at a wave period of

$$\begin{aligned}T_{(\max)} &= \left(\frac{3}{4B}\right)^{1/4} \\ &= \frac{2.00\pi V}{g} \\ &= \frac{2.00\pi(33.8)}{32.2} \\ &= 6.60 \text{ sec}\end{aligned} \tag{2.43}$$

The *maximum value of the wave spectrum* is obtained by combining equations (2.39), (2.40), (2.41), and (2.43); thus

$$S_T(T_{(\max)}) = A\left(\frac{3}{4B}\right)^{3/4} e^{-3/4}$$

$$= 1.01A\left(\frac{2\pi V}{g}\right)^3 e^{-3/4}$$

$$= 3.54 \times 10^{-3} A V^3$$

$$= 1.91 \times 10^{-5} V^3$$

$$= 0.732 \text{ ft}^2/\text{sec } (0.0679 \text{ m}^2/\text{sec}) \tag{2.44}$$

The energy per unit surface area that corresponds to the 20-knot wind velocity is

$$\bar{E} = \frac{\rho g A}{4B}$$

$$= 2.74 \times 10^{-3} \frac{\rho V^4}{g}$$

$$= 2.74 \times 10^{-3} \frac{(2.00)(33.8)^4}{32.2}$$

$$= 222 \text{ ft-lb/ft}^2 \text{ } (3220 \text{ N-m/m}^2)$$

from the results in equation (2.42). Thus a wave energy converter occupying a surface area of 100 ft^2 (9.29 m^2) will have 22,200 ft-lb (29,900 N-m) of wave energy available for conversion.

Wave data are normally presented in terms of the *significant wave*, which is defined as the average of the one-third highest waves. The *wave height* of the significant wave H_s has been found to closely agree with those heights observed by mariners. Using the results presented in the book by Newman (1977), the significant wave height resulting from a steady wind speed V in the Pierson–Moskowitz spectrum is obtained from

$$H_s = 6.49 \times 10^{-3} V^2 \quad \text{(ft)} \tag{2.45a}$$

where V is in feet per second, or

$$H_s = 0.0213 V^2 \quad \text{(m)} \tag{2.45b}$$

where V is in meters per second.

In conjunction with the significant wave height, the average wave period $\overline{T}$ of the center spectrum is used. Again, from the results of Newman (1977), the expression for the *average wave period* is

$$\overline{T} = 0.195V \tag{2.46a}$$

where V is in feet per second and

$$\overline{T} = 0.641V \tag{2.46b}$$

where V is meters per second. The data in Chapter 4 for wind-generated seas are presented in terms of H_s and $\overline{T}$.

Example 2.10

For the 20-knot wind in Example 2.9 (33.8 ft/sec or 10.3 m/sec), the significant wave height, obtained from equation (2.45), is

$$H_s = 7.41 \text{ ft } (2.26 \text{ m})$$

The average wave period of equation (2.46) is

$$\overline{T} = 6.60 \text{ sec}$$

2.4 Summary

Most wave energy conversion devices have a natural frequency, as shown in Chapter 4. Thus these devices operate most efficiently when in resonance with a monochromatic (single-frequency) wave. In Chapter 4 the performance of a number of wave energy converters are analyzed by assuming that the design wave is *linear*. Naval architects and ocean engineers have found that this assumption is satisfactory in *deep water*. In *shallow water* the monochromatic wave becomes sharp crested with a broad trough; that is, the wave profile is nonlinear (Figure 2.5). The motions of a floating wave energy convertor are then similar to the motions of the free surface, spending much more time below than above their mean position. To be accurate, therefore, the Stokes' theory or the solitary wave theory should be used in the motion analyses, depending on the wave steepness and the relative depth. These nonlinear theories are not used in this book since the purposes of the book are to demonstrate and survey the wave energy conversion techniques and ancillary systems. The mathematics involved in the nonlinear analyses would obscure these purposes.

It has been found that wave energy conversion devices operating in a *random sea* are somewhat less efficient than when operating in a sympathetic monochromatic wave. In Chapter 4 performance curves are shown for several devices operating in random seas, although the analyses are not presented. Numerous references are presented in Chapter 4 that the researcher will find useful in analyzing random sea performance.

References

Bretschneider, C. L. (1952), "The Generation and Decay of Wind Waves in Deep Water," *Transactions of the American Geophysical Union*, Vol. 33, No. 3, pp. 381–389.

Ippen, A. T., Ed. (1966), *Estuary and Coastline Hydrodynamics*, McGraw-Hill, New York.

McCormick, M. E. (1973), *Ocean Engineering Wave Mechanics*, Wiley-Interscience, New York.

McCormick, M. E. (1978), "Wind-Wave Power Available to a Wave Energy Converter Array," *Ocean Engineering*, Vol. 5, No. 2, pp. 67–74.

Newman, J. N. (1977), *Marine Hydrodynamics*, MIT Press, Cambridge, Mass.

Pierson, W. J. (1955), "Wind Generated Gravity Waves," *Advances in Geophysics*, Vol. 2, Academic, New York.

Pierson, W. J., and Moskowitz, L. (1964), "A Proposed Spectral Form for Fully Developed Wind Seas Based on the Similarity Theory of S. A. Kitaigorodski," *Transactions, American Geophysical Union*, Vol. 35, pp. 747–757.

Pierson, W. J., and Salfi, R. E. (1976), "The Temporal and Spacial Variability of Power from Ocean Waves Along the West Coast of North America," *Proceedings, Workshop on Wave and Salinity Gradient Energy Conversion*, University of Delaware, May, Paper E.

Stokes, G. G. (1847), "On the Theory of Oscillatory Waves," *Transactions*, Cambridge Philosophical Society, Vol. 8 and Papers I, No. 197.

Stokes, G. G. (1880), Supplement to the 1847 paper under the same title, *Papers*, Cambridge Philosophical Society, I, No. 314.

U.S. Army (1973), *Shore Protection Manual*, Vol. I, Coastal Engineering Research Center, Ft. Belvoir, Va.

Wiegel, R. L. (1964), *Oceanographical Engineering*, Prentice-Hall, Englewood Cliffs, N. J.

3

Wave Modification

Water waves, like sound and light waves, can be redirected by physical objects. The ways in which the waves are redirected fall into three catagories: refraction, reflection, and diffraction. *Refraction* is the turning of the wave front by a change in the water depth. *Reflection* is the reversal of the direction of wave movement due to the wave striking a barrier. *Diffraction* is the dispersion of wave energy into quiet waters in the lee of a partial barrier. These phenomena are illustrated in the following sections. Our interest in wave modification stems from recent studies into using these phenomena in wave energy conversion.

3.1 Refraction

Consider a wave approaching a beach with straight and parallel bottom contours, as sketched in Figure 3.1, where point A on the wave front is over the first shoaling contour at time t. From the results in Section 2.1 we know that this contour depth h_0 is approximately equal to half of the deep water wavelength; that is

$$h_0 \simeq \frac{\lambda_0}{2}$$

As point A travels further toward the shore, its phase velocity c decreases with the decreasing depth. The value of c at any depth can be obtained from the results presented in Figure 2.2. Since the phase velocity decreases while the period T remains unchanged, the wavelength λ also decreases by an amount determined from the results in Figure 2.2. Thus at time t' in Figure 3.1, point A on the wave front has traveled to the position of A', whereas at time t'' the point occupies the position of A''. Points B and C and all other points on the wave front will eventually experience the same deceleration. The net result of the deceleration of the wave front is a bending of the wave front which is referred to as *refraction*.

The phenomenon of refraction obeys *Snell's law*, which is mathematically

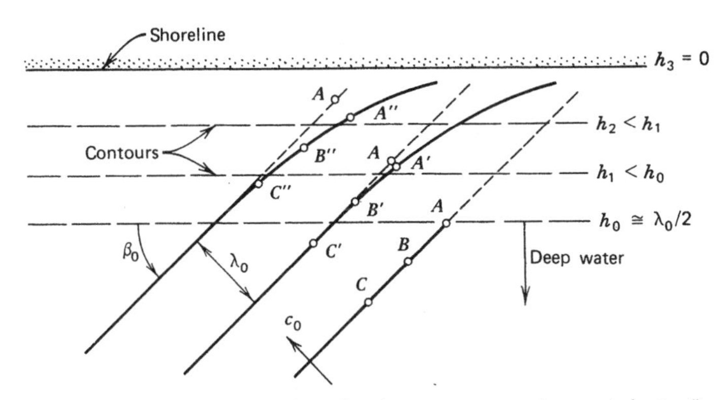

Figure 3.1 Areal sketch of a refracting wave over a time period t to t''.

defined by

$$\frac{\sin(\beta)}{\sin(\beta_0)} = \frac{c}{c_0} = \frac{\lambda}{\lambda_0} \tag{3.1}$$

where β is the angle between the wave front and the depth contour within the shoaling region and β_0 is the angle between the deep water wave front and the first shoaling contour.

The analysis of wave refraction in the *Shore Protection Manual* of the U.S. Army (1973) and elsewhere, shows that the wave height H at any point in the shoaling region is obtained from the expression

$$H = K_s K_R H_0 \tag{3.2}$$

where K_s is the shoaling coefficient, defined by

$$K_s = \left[\frac{c_{g_0}}{c_g}\right]^{1/2}$$

$$= \left\{\frac{c_0}{c\left[1 + 2kh/\sinh(2kh)\right]}\right\}^{1/2}$$

$$= \frac{H}{H_0'} \tag{3.3}$$

using the results of equations (2.20a) and (2.20b). The wave height H_0' is that of the deep water wave when no refraction occurs, that is, for $\beta_0 = 0$; thus values of the shoaling coefficient can be obtained directly from the wave height ratio in Figure 2.2. The *refraction coefficient* is denoted by K_R and is defined by

$$K_R = \left[\frac{\cos(\beta_0)}{\cos(\beta)} \right]^{1/2} \tag{3.4}$$

Refraction affects only the wave height H and the curvature of the wave front. It does not change the values of the phase velocity c or the wavelength λ at a given contour since both of these wave properties are functions of period T and depth h. For example, the values of c and λ at the contour that has a depth h_2 in Figure 3.1 will be unchanged if β_0 changes while the wave period T remains unchanged. This fact is used in the first step of the solution of refraction problems. The *steps in the analysis* are as follows:

1. The deep water values of wave height H_0, angle β_0, and wave period T are assumed to be known. The beach profile is also known.
2. At any depth contour $h < h_0$ in the shoaling region the values of λ or c are determined from the results of Figure 2.2, after determining the deep water values from equations (2.6) and (2.7), respectively.
3. The shoaling coefficient K_s is simply the value of wave height ratio in Figure 2.2.
4. The refraction angle β at the contour in question is now obtained from equation (3.1) using the results of steps 1 and 2.
5. The refraction coefficient is then determined from equation (3.4) using the results of step 4.
6. Finally the wave height H is determined from equation (3.2) using the results of steps 3 and 4.

Example 3.1

A swell that has a 7-sec period and a 3-ft (0.914-m) deep water wave height H_0 approaches a beach with straight and parallel contours. The angle between the wave front and the first shoaling contour β_0 is 45°. Our interest is in the wave height at the contour having a depth of 10 ft (3.05 m). All the properties in step 1 are known. In step 2 we need the value of the deep water wavelength λ_0. Using the results of equation (2.6),

$$\lambda_0 = \frac{gT^2}{2\pi}$$

$$= \frac{(32.2)49}{2\pi}$$

$$= 251 \text{ ft } (76.5 \text{ m})$$

The depth of the first shoaling contour is approximately

$$h_0 \simeq \frac{\lambda_0}{2}$$

$$= 126 \text{ ft } (38.3 \text{ m})$$

The h/λ_0 ratio at the contour in question is

$$\frac{h}{\lambda_0} = \frac{10}{251}$$

$$= 0.0398$$

Using this value in Figure 2.2, we obtain the following values:

$$\frac{H}{H_0'} = K_s$$

$$\simeq 1.06$$

and

$$\frac{\lambda}{\lambda_0} \simeq 0.47 = \frac{c}{c_0}$$

Using this last result in equations (3.1), we find that the angle between the 10-ft (3.05-m) contour and the wave front is

$$\beta = \sin^{-1}\left[\frac{\lambda}{\lambda_0}\sin(\beta_0)\right]$$

$$= \sin^{-1}\left[0.47 \sin(45°)\right]$$

$$= 19.4°$$

The value of the refraction coefficient, obtained from equation (3.4), is

$$K_R = \left[\frac{\cos(\beta_0)}{\cos(\beta)}\right]^{1/2}$$

$$= \left[\frac{\cos(45°)}{\cos(19.4°)}\right]^{1/2}$$

$$= 0.866$$

Finally, the wave height over the 10-ft (3.05-m) contour is obtained from equation (3.2):

$$H = K_s K_R H_0$$
$$= 1.06(0.866)3$$
$$= 2.75 \text{ ft } (0.839 \text{ m})$$

For the case of no refraction on this beach we find $\beta_0 = 0$, $K_R = 1$ and

$$H = (1.06)3$$
$$= 3.18 \text{ ft } (0.969 \text{ m})$$

Thus *refraction actually reduces the wave height from its pure shoaling value.*

Wave refraction on a beach with contours that are not straight and parallel is more difficult to analyze. In fact, the analysis is beyond the scope of this book. The reader is referred to the 1973 publication of the U.S. Army for a thorough discussion of the analytical technique.

Refraction can be used to benefit those interested in wave energy conversion. This is discussed in Section 4.2, D, where *wave focusing techniques* are analyzed.

3.2 Reflection

When a wave strikes a vertical barrier, such as a seawall, the wave energy can be *partially absorbed* by the wall if the wall is either porous or resilient. The energy not absorbed is *reflected.* Under the ideal condition of a vertical, flat, impermeable, rigid, smooth barrier in an inviscid fluid, all the incident wave energy is reflected. To analyze this latter phenomenon, consider the situation in Figure 3.2. The *incident right-running wave* causes a free-surface deflection

$$\eta^+ = \frac{H}{2}\cos(kx - \omega t) \tag{3.5}$$

assuming the wave to be linear as in equation (2.1). The wave number k is defined in equation (2.5), and the circular wave frequency ω is defined in equation (2.2). Assuming that the wave is *perfectly reflected*, the reflected wave is a *left-running wave* described by

$$\eta^- = \frac{H}{2}\cos(kx + \omega t) \tag{3.6}$$

(***Note***: The only difference in the expression for η^+ and η^- is in the sign of the time-dependent term.)

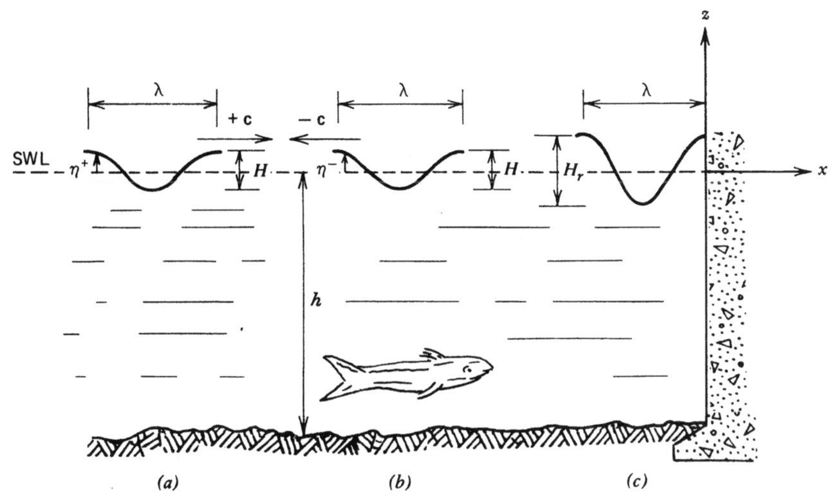

Figure 3.2 Perfect reflection of an incident wave from a smooth, vertical seawall resulting in a standing wave: (*a*) incident wave; (*b*) reflected wave; (*c*) standing wave.

The expressions of equations (3.5) and (3.6) can be added to each other because of the *superposition* property of linear waves. The result is

$$\begin{aligned} \eta &= \eta^+ + \eta^- \\ &= 2\left(\frac{H}{2}\right)\cos(kx)\cos(\omega t) \\ &= \frac{H_r}{2}\cos(kx)\cos(\omega t) \end{aligned} \tag{3.7}$$

which describes a *standing wave* whose height H_r is twice the incident wave height H. The standing wave has no phase velocity since the phase velocities of the incident and reflected waves cancel each other; however, the wavelength λ is unchanged.

The *energy* of the standing wave is the sum of the energies of the incident and reflected waves:

$$\begin{aligned} E_r &= E^+ + E^- \\ &= \frac{2\rho g H^2 \lambda b}{8} \\ &= \frac{\rho g H^2 \lambda b}{4} \end{aligned} \tag{3.8a}$$

using the results of equation (2.17). When the standing wave has its maximum deflection from the SWL, as illustrated in Figure 3.2, the energy of the wave is totally *potential energy*. Thus from equation (2.18)

$$E_r = \frac{\rho g H_r^2 \lambda b}{16} \tag{3.8b}$$

Since $H_r = 2H$, we see that the expressions in equations (3.8a) and (3.8b) are identical. The significance of the result of equation (3.8a) in wave energy conversion is obvious. If a reflecting barrier is placed in the lee of a wave energy conversion device, *the energy available to the device is doubled* under ideal conditions.

Example 3.2

The swell in Example 3.1 has a height of 3 ft (0.914 m) and a period of 7 sec in deep water. It strikes a vertical flat barrier located in 10 ft (3.05 m) of water and is perfectly reflected. To determine the standing wave properties at the barrier, we must first determine those of the incident waves in the 10 ft (3.05-m) depth. To do this, we use the results presented in the shoaling diagrams in Figure 2.2. To enter Figure 2.2, the value of the deep water wavelength λ_0 is required. Using the results of Example 3.1, $\lambda_0 = 251$ ft (76.5 m). Thus in Figure 2.2 $h/\lambda_0 = 0.0398$,

$$\frac{\lambda}{\lambda_0} = 0.47$$

and

$$\frac{H}{H_0} = 1.06$$

The values of the wavelength and the wave height at the barrier are, respectively, 118 ft (36.0 m) and 3.18 ft (0.969 m).

The expression for the free-surface profile of the standing wave at the barrier is obtained from equation (3.7); thus

$$\eta_r = 2\left(\frac{H}{2}\right)\cos(kx)\cos(\omega t)$$

$$= 2\left(\frac{3}{2}\right)\cos\left(\frac{2\pi x}{118}\right)\cos\left(\frac{2\pi t}{7}\right)$$

$$= 3\cos(0.0532x)\cos(0.898t)$$

for x in feet and

$$\eta_r = 0.914\cos(0.174x)\cos(0.898t)$$

where x is in meters. The height of the wave at the barrier H_r is twice that of the incident wave; that is,

$$H_r = 2H$$

$$= 6 \text{ ft } (1.83 \text{ m})$$

The energy per unit crest width of the standing wave, from equation (3.8b), is

$$\frac{E_r}{b} = \frac{\rho g H_r^2 \lambda}{16}$$

$$= \frac{2.00(32.2)6^2(118)}{16}$$

$$= 17{,}100 \text{ ft-lb/ft}$$

or

$$\frac{E_r}{b} = 76{,}000 \text{ N-m/m}$$

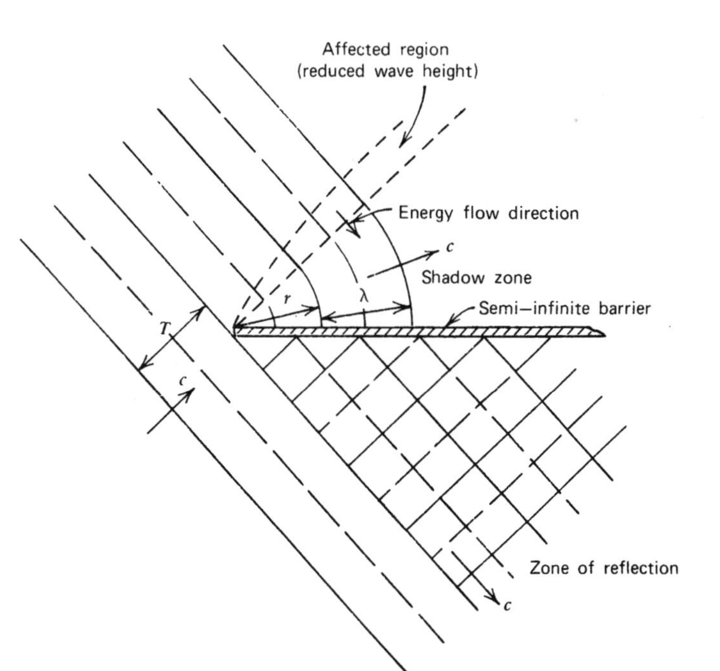

Figure 3.3 Wave diffraction and oblique wave reflection from a semi-infinite, thin, vertical barrier.

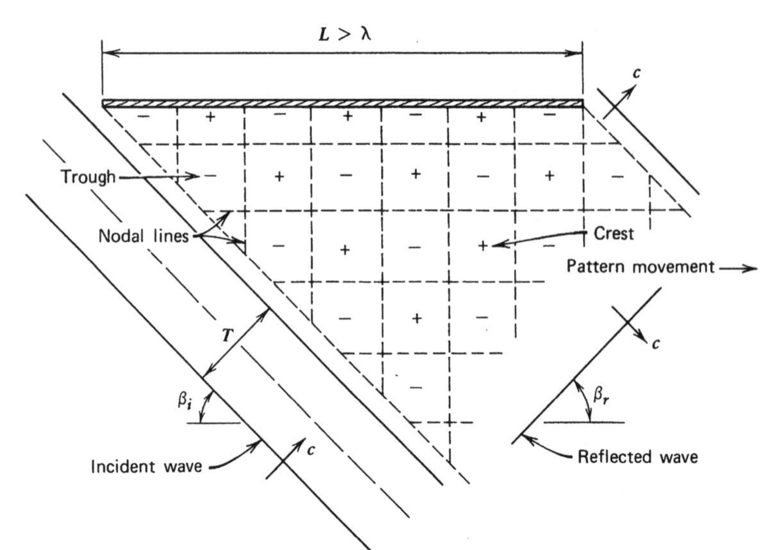

Figure 3.4 Perfectly reflected oblique wave pattern from a finite vertical barrier.

Those readers interested in the forces and moments induced by standing waves should consult the books of McCormick (1973) or Wiegel (1964) or that edited by Ippen (1966).

The phenomenon of *oblique wave reflection* is illustrated in Figure 3.3 along with that of diffraction. For *perfect reflection* of oblique waves the wave pattern resembles an undulating "checkerboard" that is convected along the wall as illustrated in Figure 3.4, where the angle of incident β_i is equal to the angle of reflection β_r. In Figure 3.3 a *crest* occurs in the *zone of reflection* when crests from the incident and reflected wave patterns cross each other; these are denoted by (+) in Figure 3.4. When incident and reflected wave troughs coincide, a *trough* occurs in the zone of reflection and is denoted by (−) in Figure 3.4. Finally, when an incident crest and a reflected trough (or vice versa) coincide, a *node* results. In this case there are *nodal lines* that separate the troughs and the crests, as illustrated in Figure 3.4. Oblique wave reflection is also discussed in the book by Svendsen and Jonsson (1976). For the designer of a wave energy conversion system, the ideal condition is that when the incident wave angle β_i is zero, since the crest–trough positions are then fixed.

3.3 Diffraction

Diffraction is the passage of wave energy into the calm waters in the lee of obstacle. The calm water region is called the *shadow zone* since the incident wave cannot "see" this region. As illustrated in Figure 3.3, when the incident wave clears the barrier, the wave front becomes *circular* in the shadow zone and appears to radiate from the end of the barrier. The phenomenon is caused by nature's attempt to fill an energy void. Since energy is flowing into the shadow zone in a direction parallel to the incident wave crest, the wave height in the shadow zone H_D is less than the incident wave height H. In addition, there is a decrease in the wave height just outside the shadow zone since this is the region supplying the wave energy to the shadow zone. If the water depth is constant, the wavelengths and the phase velocities of the incident and diffracted waves are equal.

The analysis of diffraction is beyond the scope of this book. For those readers interested in obtaining information on the analytical technique, the paper by Wiegel (1962) is recommended. For our purposes, results of the analysis and associated experiments will suffice. These results are presented in Table 3.1, in Wiegel (1962), and in Figures 3.5 through 3.7 from the 1973 U.S. Army publication. In Figure 3.5 we see the dimensionless values of the diffracted wave heights as functions of position for an approach angle of 45° for a semi-infinite barrier. A constant water depth h is assumed in this plot. The ratio of the wave heights is called the *diffraction coeffient* and is denoted by

$$K_D = \frac{H_D}{H} \tag{3.9}$$

In Figure 3.6 the approach angle is 90°, whereas in Figure 3.7 the approach angle is 135° for the semi-infinite barrier in waters of constant depth. In the case of a *finite-length barrier* of length $L \gg \lambda$, the diffraction patterns from the two ends of the barriers can be *superimposed* on each other to obtain the wave height H_D in the shadow zone.

Example 3.3

An incident wave of 3-ft (0.914-m) height and 7-sec period strikes the semi-infinite vertical flat barrier described in Example 3.2 at an angle of $\alpha = 45°$. The water depth is 10 ft (3.05 m). From Example 3.2, the wavelength is

$$\lambda = 118 \text{ ft } (36.0 \text{ m})$$

The wave height H_D in the shadow zone at a position 590 ft (180 m) from the end of the barrier at an angle $\theta = 15°$ from the barrier can be obtained from the results of the matrix in Table 3.1. Entering that matrix with the values of $\alpha = 45°$, $\theta = 15°$, and $r/\lambda = 5$, we find the diffraction coefficient to be

$$K_D = \frac{H_D}{H}$$

$$= 0.20$$

Table 3.1 Wave Diffraction Coefficients, $K_D = H_D/H$

	θ (deg)												
r/λ	0	15	30	45	60	75	90	105	120	135	150	165	180
						$\alpha = 15°$							
$\frac{1}{2}$	0.49	0.79	0.83	0.90	0.97	1.01	1.03	1.02	1.01	0.99	0.99	1.00	1.00
1	0.38	0.73	0.83	0.95	1.04	1.04	0.99	0.98	1.01	1.01	1.00	1.00	1.00
2	0.21	0.68	0.86	1.05	1.03	0.97	1.02	0.99	1.00	1.00	1.00	1.00	1.00
5	0.13	0.63	0.99	1.04	1.03	1.02	0.99	0.99	1.00	1.01	1.00	1.00	1.00
10	0.35	0.58	1.10	1.05	0.98	0.99	1.01	1.00	1.00	1.00	1.00	1.00	1.00
						$\alpha = 30°$							
$\frac{1}{2}$	0.61	0.63	0.68	0.76	0.87	0.97	1.03	1.05	1.03	1.01	0.99	0.95	1.00
1	0.50	0.53	0.63	0.78	0.95	1.06	1.05	0.98	0.98	1.01	1.01	0.97	1.00
2	0.40	0.44	0.59	0.84	1.07	1.03	0.96	1.02	0.98	1.01	0.99	0.95	1.00
5	0.27	0.32	0.55	1.00	1.04	1.04	1.02	0.99	0.99	1.00	1.01	0.97	1.00
10	0.20	0.24	0.54	1.12	1.06	0.97	0.99	1.01	1.00	1.00	1.00	0.98	1.00
						$\alpha = 45°$							
$\frac{1}{2}$	0.49	0.50	0.55	0.63	0.73	0.85	0.96	1.04	1.06	1.04	1.00	0.99	1.00
1	0.38	0.40	0.47	0.59	0.76	0.95	1.07	1.06	0.98	0.97	1.01	1.01	1.00
2	0.29	0.31	0.39	0.56	0.83	1.08	1.04	0.96	1.03	0.98	1.01	1.00	1.00
5	0.18	0.20	0.29	0.54	1.01	1.04	1.05	1.03	1.00	0.99	1.01	1.00	1.00
10	0.13	0.15	0.22	0.53	1.13	1.07	0.96	0.98	1.02	0.99	1.00	1.00	1.00
						$\alpha = 60°$							
$\frac{1}{2}$	0.40	0.41	0.45	0.52	0.60	0.72	0.85	1.13	1.04	1.06	1.03	1.01	1.00
1	0.31	0.32	0.36	0.44	0.57	0.75	0.96	1.08	1.06	0.98	0.98	1.01	1.00
2	0.22	0.23	0.28	0.37	0.55	0.83	1.08	1.04	0.96	1.03	0.98	1.01	1.00
5	0.14	0.15	0.18	0.28	0.53	1.01	1.04	1.05	1.03	0.99	0.99	1.00	1.00
10	0.10	0.11	0.13	0.21	0.52	1.14	1.07	0.96	0.98	1.01	1.00	1.00	1.00
						$\alpha = 75°$							
$\frac{1}{2}$	0.34	0.35	0.38	0.42	0.50	0.59	0.71	0.85	0.97	1.04	1.05	1.02	1.00
1	0.25	0.26	0.29	0.34	0.43	0.56	0.75	0.95	1.02	1.06	0.98	0.98	1.00
2	0.18	0.19	0.22	0.26	0.36	0.54	0.83	1.09	1.04	0.96	1.03	0.99	1.00
5	0.12	0.12	0.13	0.17	0.27	0.52	1.01	1.04	1.05	1.03	0.99	0.99	1.00
10	0.08	0.08	0.10	0.13	0.20	0.52	1.14	1.04	0.96	0.98	1.01	1.00	1.00
						$\alpha = 90°$							
$\frac{1}{2}$	0.31	0.31	0.33	0.36	0.41	0.49	0.59	0.71	0.85	0.96	1.03	1.03	1.00
1	0.22	0.23	0.24	0.28	0.33	0.42	0.56	0.75	0.96	1.07	1.05	0.99	1.00
2	0.16	0.16	0.18	0.20	0.26	0.35	0.54	0.69	1.08	1.04	0.96	1.02	1.00
5	0.10	0.10	0.11	0.13	0.16	0.27	0.53	1.01	1.04	1.05	1.02	0.99	1.00
10	0.07	0.07	0.08	0.09	0.13	0.20	0.52	1.14	1.07	0.96	0.99	1.01	1.00
						$\alpha = 105°$							
$\frac{1}{2}$	0.28	0.28	0.29	0.32	0.35	0.41	0.49	0.59	0.72	0.85	0.97	1.01	1.00
1	0.20	0.20	0.21	0.23	0.27	0.33	0.42	0.56	0.75	0.95	1.06	1.04	1.00
2	0.14	0.14	0.13	0.17	0.20	0.25	0.35	0.54	0.83	1.08	1.03	0.97	1.00
5	0.09	0.09	0.10	0.11	0.13	0.17	0.27	0.52	1.02	1.04	1.04	1.02	1.00
10	0.07	0.06	0.08	0.08	0.09	0.12	0.20	0.52	1.14	1.07	0.97	0.99	1.00

Table 3.1 (*Continued*)

r/λ	θ (deg) 0	15	30	45	60	75	90	105	120	135	150	165	180
						$\alpha = 120°$							
$\frac{1}{2}$	0.25	0.26	0.27	0.28	0.31	0.35	0.41	0.50	0.60	0.73	0.87	0.97	1.00
1	0.18	0.19	0.19	0.21	0.23	0.27	0.33	0.43	0.57	0.76	0.95	1.04	1.00
2	0.13	0.13	0.14	0.14	0.17	0.20	0.26	0.36	0.55	0.83	1.07	1.03	1.00
5	0.08	0.08	0.08	0.09	0.11	0.13	0.16	0.27	0.53	1.01	1.04	1.03	1.00
10	0.06	0.06	0.06	0.07	0.07	0.09	0.13	0.20	0.52	1.13	1.06	0.98	1.00
						$\alpha = 135°$							
$\frac{1}{2}$	0.24	0.24	0.25	0.26	0.28	0.32	0.36	0.42	0.52	0.63	0.76	0.90	1.00
1	0.18	0.17	0.18	0.19	0.21	0.23	0.28	0.34	0.44	0.59	0.78	0.95	1.00
2	0.12	0.12	0.13	0.14	0.14	0.17	0.20	0.26	0.37	0.56	0.84	1.05	1.00
5	0.08	0.07	0.08	0.08	0.09	0.11	0.13	0.17	0.28	0.54	1.00	1.04	1.00
10	0.05	0.06	0.06	0.06	0.07	0.08	0.09	0.13	0.21	0.53	1.12	1.05	1.00
						$\alpha = 150°$							
$\frac{1}{2}$	0.23	0.23	0.24	0.25	0.27	0.29	0.33	0.38	0.45	0.55	0.68	0.83	1.00
1	0.16	0.17	0.17	0.18	0.19	0.22	0.24	0.29	0.36	0.47	0.63	0.83	1.00
2	0.12	0.12	0.12	0.13	0.14	0.15	0.18	0.22	0.28	0.39	0.59	0.86	1.00
5	0.07	0.07	0.08	0.08	0.08	0.10	0.11	0.13	0.18	0.29	0.55	0.99	1.00
10	0.05	0.05	0.05	0.06	0.06	0.07	0.08	0.10	0.13	0.22	0.54	1.10	1.00
						$\alpha = 165°$							
$\frac{1}{2}$	0.23	0.23	0.23	0.24	0.26	0.28	0.31	0.35	0.41	0.50	0.63	0.79	1.00
1	0.16	0.16	0.17	0.17	0.19	0.20	0.23	0.26	0.32	0.40	0.53	0.73	1.00
2	0.11	0.11	0.12	0.12	0.13	0.14	0.16	0.19	0.23	0.31	0.44	0.68	1.00
5	0.07	0.07	0.07	0.07	0.08	0.09	0.10	0.12	0.15	0.20	0.32	0.63	1.00
10	0.05	0.05	0.05	0.06	0.06	0.06	0.07	0.08	0.11	0.11	0.21	0.58	1.00
						$\alpha = 180°$							
$\frac{1}{2}$	0.20	0.25	0.23	0.24	0.25	0.28	0.31	0.34	0.40	0.49	0.61	0.78	1.00
1	0.10	0.17	0.16	0.18	0.18	0.23	0.22	0.25	0.31	0.38	0.50	0.70	1.00
2	0.02	0.09	0.12	0.12	0.13	0.18	0.16	0.18	0.22	0.29	0.40	0.60	1.00
5	0.02	0.06	0.07	0.07	0.07	0.08	0.10	0.12	0.14	0.18	0.27	0.46	1.00
10	0.01	0.05	0.05	0.04	0.06	0.07	0.07	0.08	0.10	0.13	0.20	0.36	1.00

Source: Wiegel (1962).

Thus the wave height at this location is

$$H_D = 0.60 \text{ ft } (0.183 \text{ m})$$

Example 3.4

The wave condition in Example 3.3 is applied to a finite barrier of length $L = 708$ ft (2.16 m). Since $L/\lambda = 6$ we can superimpose the diffraction patterns resulting from the two ends of the barrier, that is, the respective

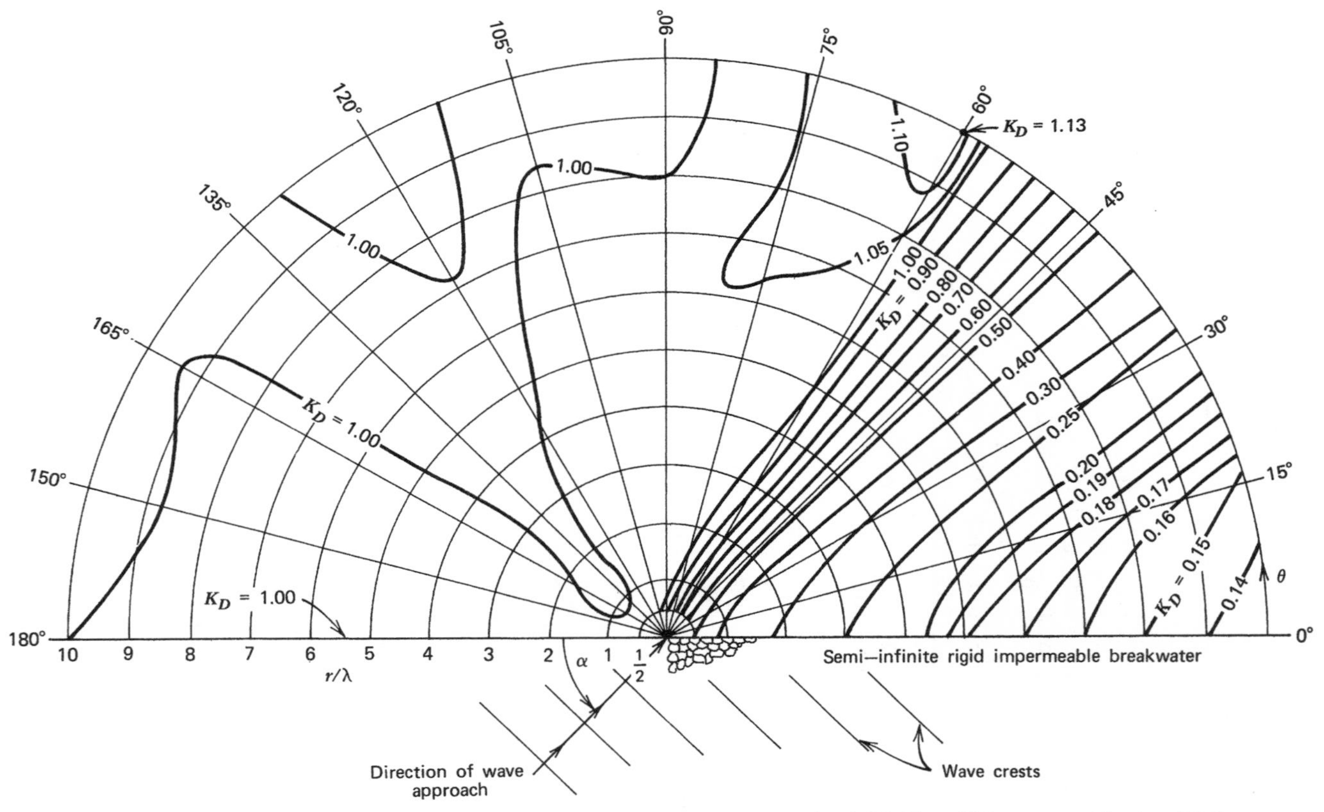

Figure 3.5 Diffraction from a semi-infinite barrier with an incident wave angle of 45°. From U.S. Army publication (1973).

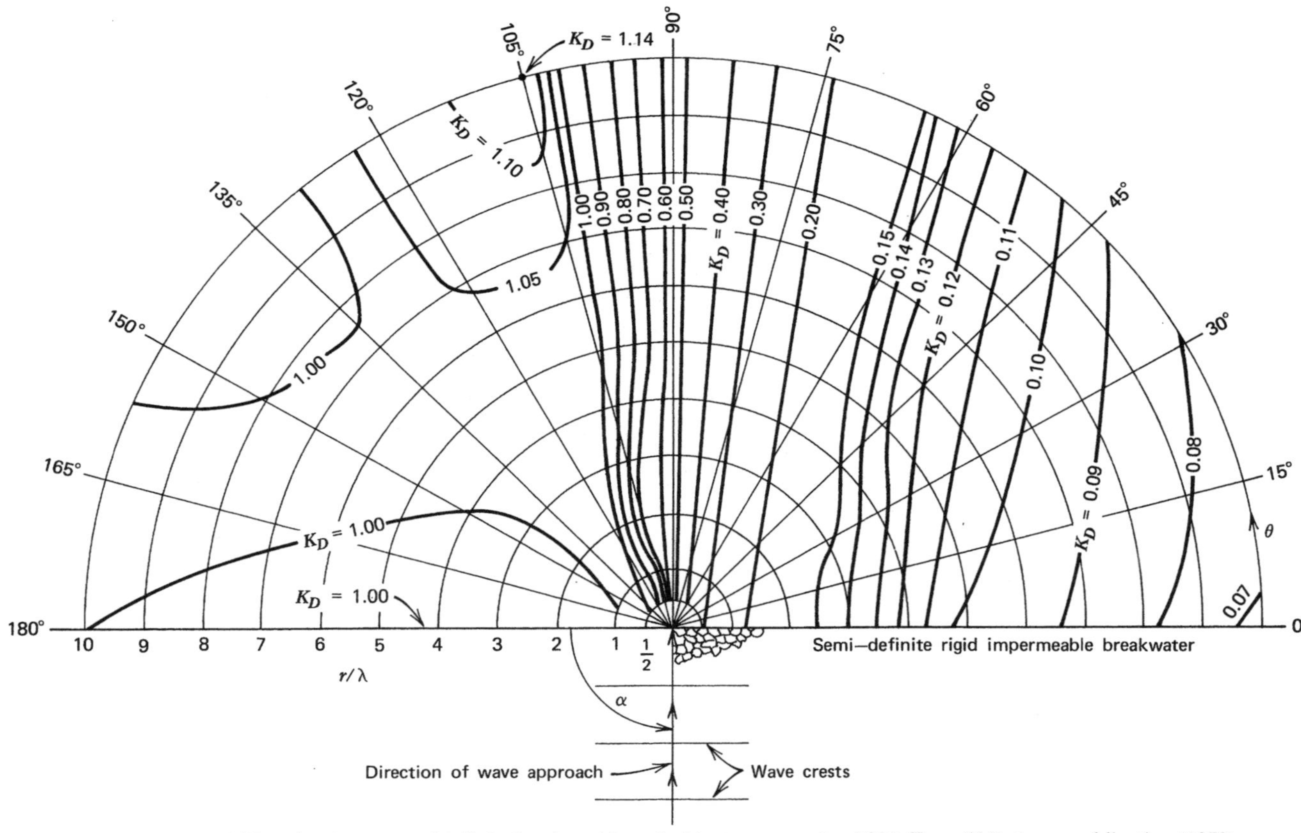

Figure 3.6 Diffraction from a semi-infinite barrier with an incident wave angle of 90°. From U.S. Army publication (1973).

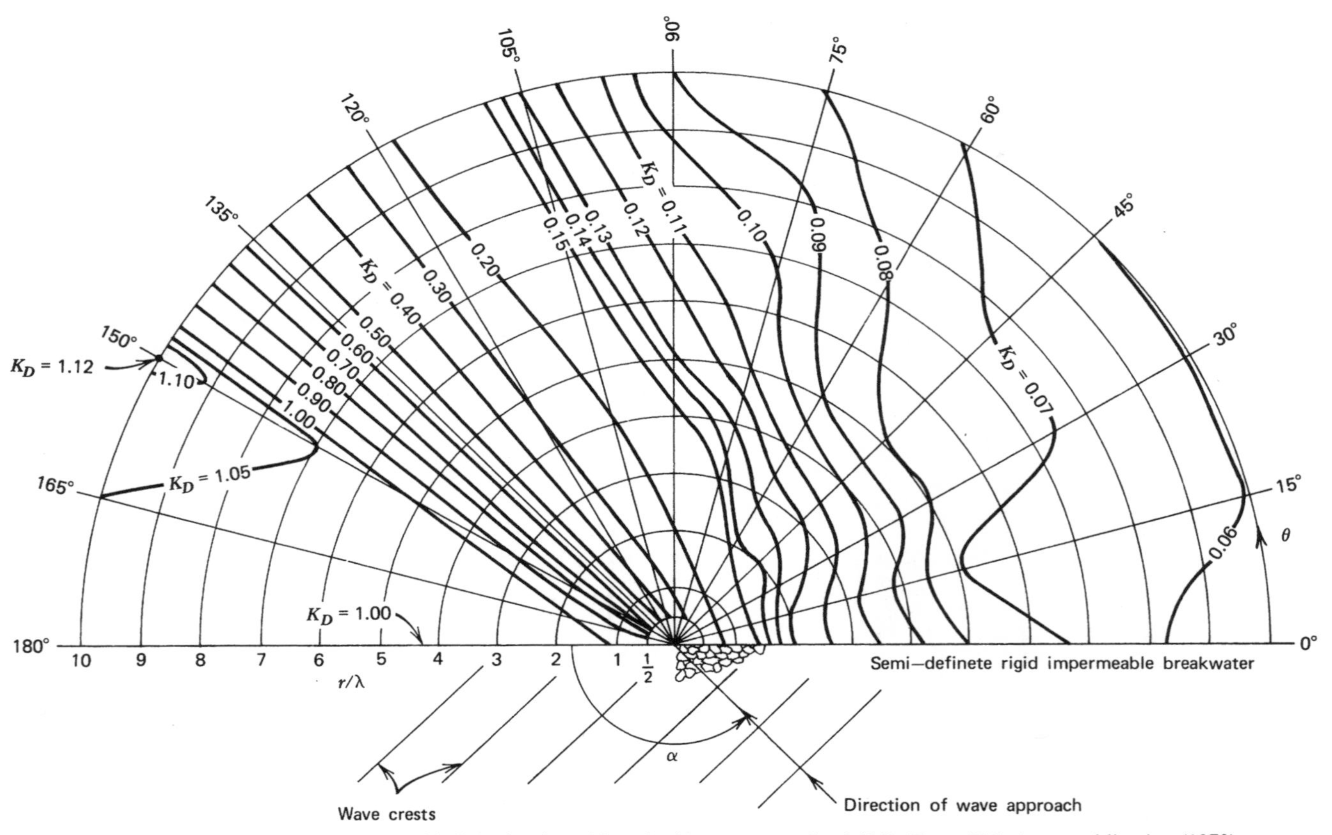

Figure 3.7 Diffraction from a semi-infinite barrier with an incident wave angle of 135°. From U.S. Army publication (1973).

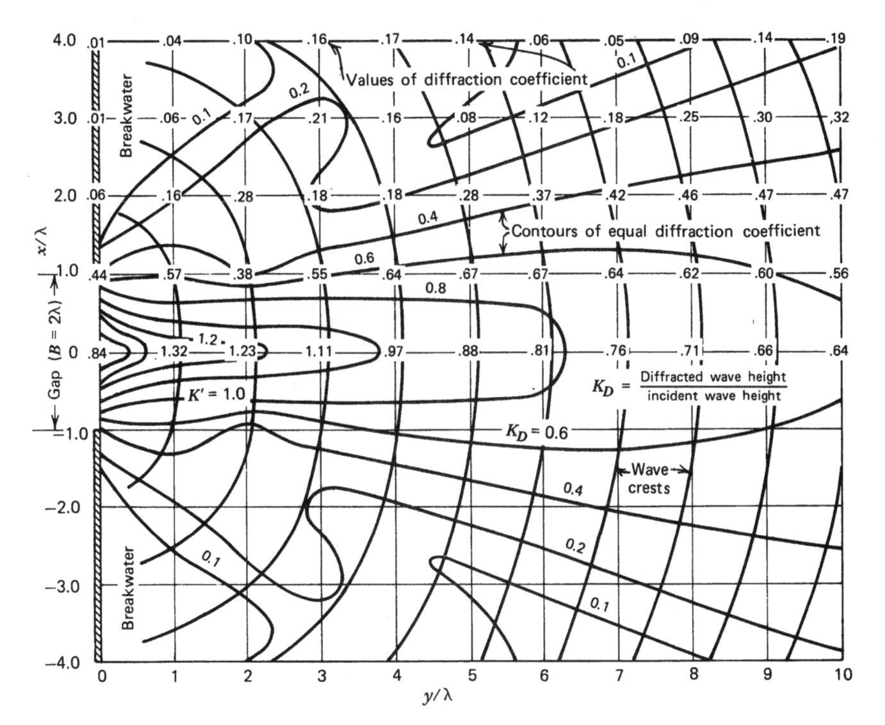

Figure 3.8 Diffraction from a gap with an incident wave angle of 90°. From U.S. Army publication (1973).

results of Figures 3.5 and 3.7. At a point midway between the two ends and on the lee side of the barrier, $r/\lambda = 3$ in both Figures 3.5 and 3.7, whereas $\theta = 0°$ in those figures. The respective diffraction coefficients from Figures 3.5 and 3.7 are then

$$K'_D = 0.250$$

and

$$K''_D = 0.105$$

Thus the wave height at the midpoint of the lee side of the barrier is

$$\begin{aligned} H_D &= H(K'_D + K''_D) \\ &= 3(0.250 + 0.105) \\ &= 1.06 \text{ ft } (0.349 \text{ m}) \end{aligned}$$

The use of finite barriers in wave energy conversion is for wave *control* rather than energy conversion. Mehlum and Stamnes (1979) have illustrated that diffraction through a *gap* can be used to *focus* wave energy on a point where a wave energy conversion device may be located. Significant work on gap diffraction of water waves was performed by Johnson (1953) and Wiegel (1964). Using the results of Johnson (1953) as presented in the 1973 U.S. Army publication, the diffraction pattern from a gap of width B equal to twice the incident wavelength is presented in Figure 3.8, where $\alpha = 90°$. In the case of $B \gg \lambda$ the diffraction pattern is approximately that of Figure 3.6, which is applied to each side of the gap. For the case of the gap width of the same order of magnitude as the wavelength and an incident wave angle α less than 90°, the reader is referred to Chapter 8 of Wiegel's (1964) book. This is the case needed to observe the focusing reported by Mehlum and Stamnes (1979).

References

Ippen, A. T., Ed. (1966), *Estuary and Coastline Hydrodynamics*, McGraw-Hill, New York.

Johnson, J. W. (1953), "Engineering Aspects of Diffraction and Wave Refraction," *Transactions American Society of Civil Engineers*, Vol. 118, No. 2556, pp. 617–652.

McCormick, M. E. (1973), *Ocean Engineering Wave Mechanics*, Wiley-Interscience, New York.

Mehlum, E., and Stamnes, J. (1979), "On the Focusing of Ocean Swells and Its Significance in Power Production," *Proceedings, Symposium on Wave Energy Utilization*, Chalmers University, Gothenburg, Sweden.

Svendsen, I. A., and Jonsson, I. G. (1976), *Hydrodynamics of Coastal Region*, Technical University of Denmark, Lyngby.

U.S. Army (1973), *Shore Protection Manual*, Fort Belvoir, Va.

Wiegel, R. L. (1962), "Diffraction of Waves by a Semi-Infinite Breakwater," *Journal of the Hydraulics Division*, American Society of Civil Engineers, Vol. 88, HY1, January, pp. 27–44.

Wiegel, R. L. (1964), *Oceanographical Engineering*, Prentice-Hall, Englewood Cliffs, N. J.

4

Wave Energy Conversion

The idea of converting the energy of water waves into useful energy forms is not new. There are techniques that were first patented in the nineteenth century and, in addition, references in the technical literature to ideas that predate these techniques. Some of the wave energy conversion patents are listed in Appendix B.

When one observes the waves generated by a storm at sea, one is impressed with the enormous power of the storm waves. Waves that can lift the bow of a 20,000-ton ship out of the water have excited the imaginations of many inventors over the centuries. These inventors have, in turn, devised methods of exploiting the waves by converting the wave energy into mechanical and electrical energy, some with a fair degree of success. For example, the Bouchaux-Praceique wave energy convertor, sketched in Figure 1.5, was operational in the early twentieth century, according to Palme (1920).

Although there are over 1000 patented wave energy conversion techniques in Japan, North America, and Europe, there are only nine basic ideas on which these techniques are based. In this chapter these ideas are described and several examples of each presented.

4.1 Basic Wave Energy Conversion Techniques

We begin our discussion by stating a basic principle, namely, that any technique that can effectively create waves can also be used to extract wave energy. A number of experimental naval architects and ocean engineers have suggested using hinged vertical wave boards, moving bulkheads, plungers, and pneumatic chambers as wave energy converters since these devices are effective in creating waves. Some of these wave energy conversion ideas were studied under experimental conditions and were found to work with moderate efficiency under rather specific conditions. These conditions involve the frequency of the monochromatic waves and the relative alignment of the wave front and the face of the device in question. In the following sections some of these basic techniques are described and discussed.

A Heaving and Pitching Bodies

Consider the wave energy conversion schemes sketched in Figure 4.1. In Figure 4.1*a* the body is allowed to *heave* only; that is, the motion is in a vertical direction. The heaving motion of the float excites some electromechanical energy conversion device (not shown). This device might be similar to those described in Section 5.2.

Let us first consider the effects of the relative lengths of the body L and the wave λ. When $\lambda = L$, assuming a *sinusoidal wave profile* from equation

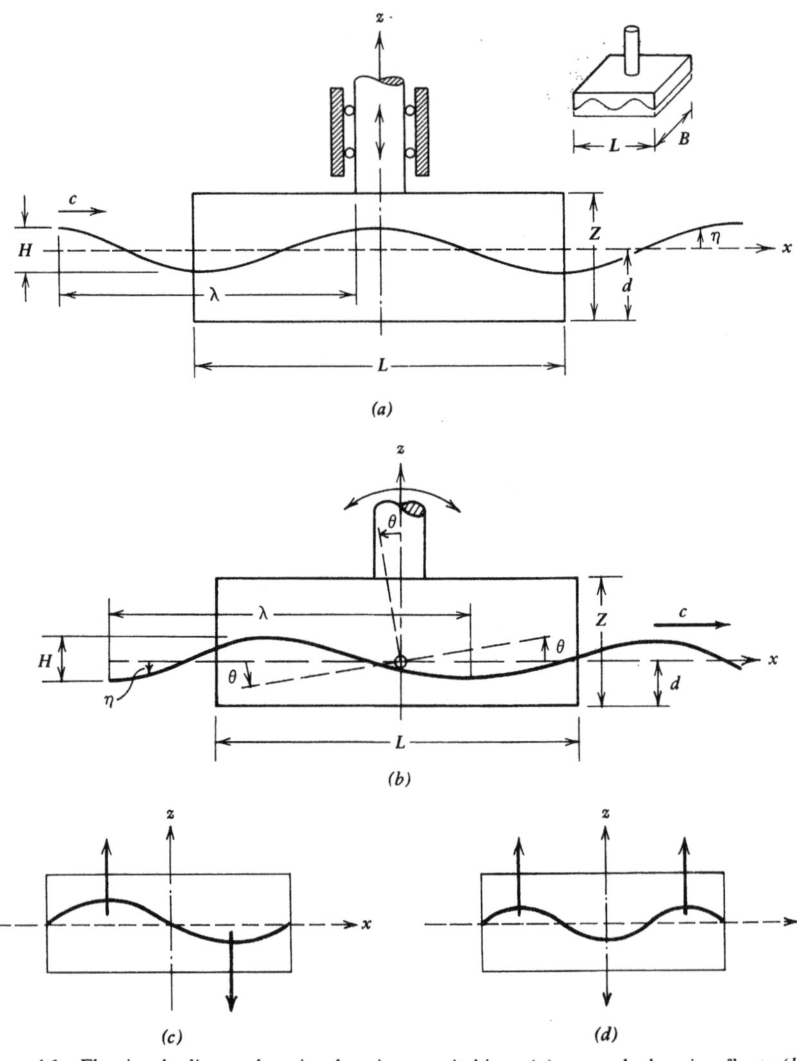

Figure 4.1 Floating bodies undergoing heaving or pitching: (*a*) a purely heaving float; (*b*) a purely pitching float; (*c*) pure pitching condition; (*d*) pure heaving condition.

(2.1), no heaving motion will occur since both a crest and a trough of the wave occur simultaneously over the body length as illustrated in Figure 4.1*c*. From the discussion of waves in Chapter 2, we see that the pressure forces due to the water particle motions near two successive nodes (where $\eta = 0$) cancel each other. Furthermore, the increased buoyant force due to a crest is canceled by the decreased buoyancy due to the trough. Following this logic we can conclude that there is *no net vertical force* on the float when

$$L = N\lambda, \qquad N = 1, 2, 3, \ldots \tag{4.1}$$

and, therefore, *no heaving motions*. Conversely, when an "extra" trough or crest occurs over the body length as illustrated in Figure 4.1*d*, then there is a *net vertical force*. Heaving can be expected when

$$L = \frac{N\lambda}{2}, \qquad N = 1, 3, 5, \ldots \tag{4.2}$$

The body sketched in Figure 4.1*b* is allowed only to *pitch* (rotate) about the center of gravity. A *minimum pitching moment* is experienced in a monochromatic wave of length λ under the condition described by equation (4.2) when $N = 3$. On the other hand, a *maximum pitching moment* occurs when equation (4.1) is satisfied, such as in Figure 4.1*c*, where $N = 1$.

In a *random sea*, discussed in Section 2.3, a wave energy converter is subject to waves of varying heights and periods. Thus the conditions described in equations (4.1) and (4.2) are seldom encountered in a wind-generated sea. When the waves have traveled great distances from the storm region, *dispersion* has taken place so that the long waves (low period) have outraced the short waves. Then the waves encountered are *swells* and are nearly monochromatic.

A floating system such as those shown in Figure 4.1 will have either a natural heaving period or a natural pitching period or both if the float is unconstrained. The wave energy conversion designer then tries to design one or more of these natural frequncies to *resonate* with either the highest energy wave in a wind generated sea (see Section 2.3) or a predominant swell. As derived by McCormick (1973), Bhattacharyya (1978), and others, the *natural heaving frequency* of a floating body is

$$f_z = \frac{1}{T_z} = \frac{\omega_z}{2\pi} = \frac{1}{2\pi}\sqrt{\frac{\rho g A_{wp}}{m + m_w}} \tag{4.3}$$

where T_z is the natural heaving period, ω_z is the natural circular heaving frequency, ρ is the mass density of seawater (2.00 slugs/ft^3, 1,030 kg/m^3), A_{wp} is the waterplane area of the float, m is the mass of the heaving system, and m_w is the added mass, that is, the mass of the water excited by the heaving motion. Expressions for the added mass are presented in Figures 4.2 and 4.3. For the purpose of this section, we confine our attention to two basic shapes:

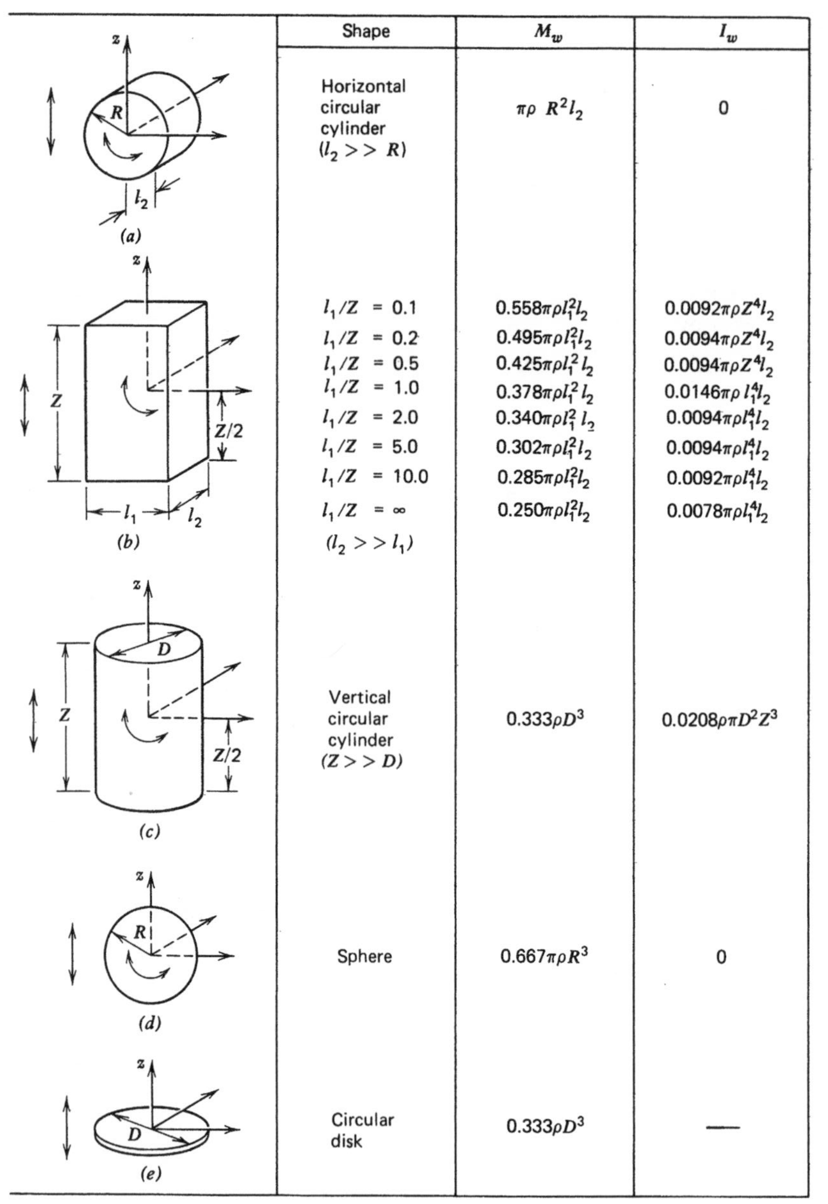

Shape	M_w	I_w
Horizontal circular cylinder ($l_2 >> R$)	$\pi\rho\ R^2 l_2$	0
$l_1/Z = 0.1$	$0.558\pi\rho l_1^2 l_2$	$0.0092\pi\rho Z^4 l_2$
$l_1/Z = 0.2$	$0.495\pi\rho l_1^2 l_2$	$0.0094\pi\rho Z^4 l_2$
$l_1/Z = 0.5$	$0.425\pi\rho l_1^2 l_2$	$0.0094\pi\rho Z^4 l_2$
$l_1/Z = 1.0$	$0.378\pi\rho l_1^2 l_2$	$0.0146\pi\rho l_1^4 l_2$
$l_1/Z = 2.0$	$0.340\pi\rho l_1^2 l_2$	$0.0094\pi\rho l_1^4 l_2$
$l_1/Z = 5.0$	$0.302\pi\rho l_1^2 l_2$	$0.0094\pi\rho l_1^4 l_2$
$l_1/Z = 10.0$	$0.285\pi\rho l_1^2 l_2$	$0.0092\pi\rho l_1^4 l_2$
$l_1/Z = \infty$ ($l_2 >> l_1$)	$0.250\pi\rho l_1^2 l_2$	$0.0078\pi\rho l_1^4 l_2$
Vertical circular cylinder ($Z >> D$)	$0.333\rho D^3$	$0.0208\rho\pi D^2 Z^3$
Sphere	$0.667\pi\rho R^3$	0
Circular disk	$0.333\rho D^3$	—

Figure 4.2 Added-mass (m_w) and added-mass moment of inertia (I_w) expressions for fully submerged bodies. From Wendel (1956).

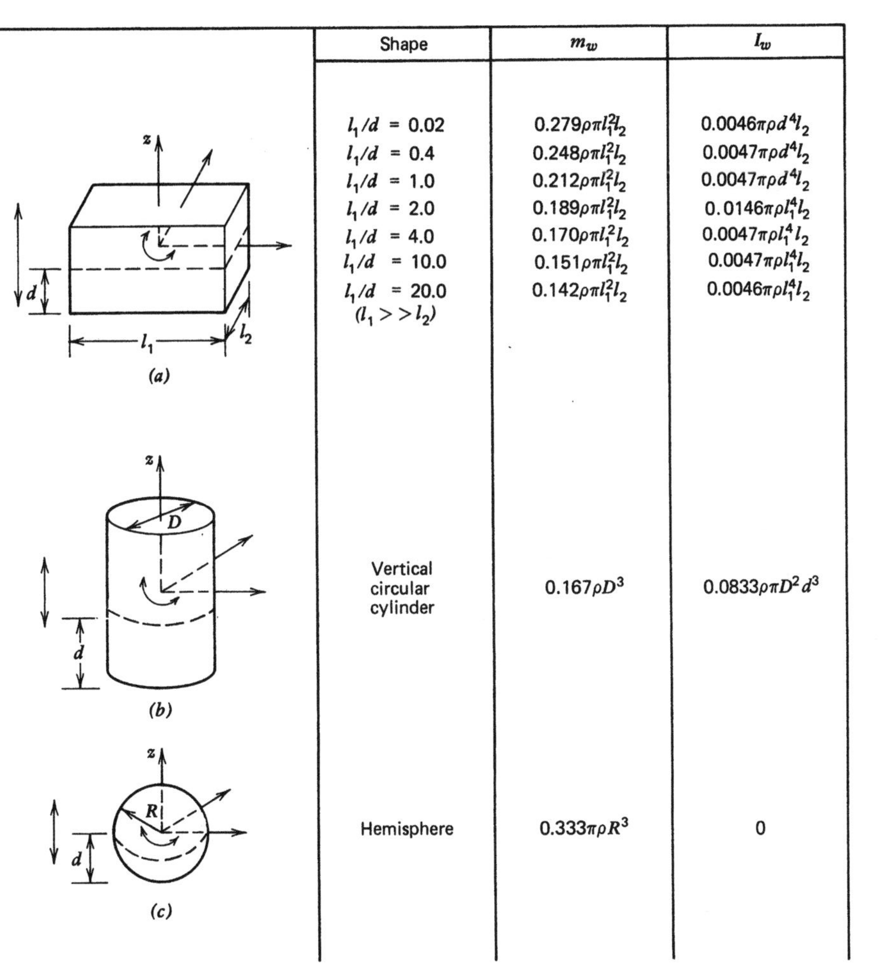

Shape	m_w	I_w
$l_1/d = 0.02$	$0.279\rho\pi l_1^2 l_2$	$0.0046\pi\rho d^4 l_2$
$l_1/d = 0.4$	$0.248\rho\pi l_1^2 l_2$	$0.0047\pi\rho d^4 l_2$
$l_1/d = 1.0$	$0.212\rho\pi l_1^2 l_2$	$0.0047\pi\rho d^4 l_2$
$l_1/d = 2.0$	$0.189\rho\pi l_1^2 l_2$	$0.0146\pi\rho l_1^4 l_2$
$l_1/d = 4.0$	$0.170\rho\pi l_1^2 l_2$	$0.0047\pi\rho l_1^4 l_2$
$l_1/d = 10.0$	$0.151\rho\pi l_1^2 l_2$	$0.0047\pi\rho l_1^4 l_2$
$l_1/d = 20.0$ $(l_1 >> l_2)$	$0.142\rho\pi l_1^2 l_2$	$0.0046\pi\rho l_1^4 l_2$
Vertical circular cylinder	$0.167\rho D^3$	$0.0833\rho\pi D^2 d^3$
Hemisphere	$0.333\pi\rho R^3$	0

Figure 4.3 Added-mass (m_w) and added-mass moment of inertia (I_w) expressions for half-submerged bodies: (*a*) from Wendel (1956); (*b*) from Hooft (1970); (*c*) from Lamb (1932).

the floating rectangular solid (Figure 4.3*a*) and the floating circular cylinder with a vertical axis (Figure 4.3*b*). The theoretical added-mass expressions for these two shapes are the following:

$$\text{Rectangular solid:} \qquad m_w = \frac{K_m \pi \rho L B^2}{4} \tag{4.4}$$

$$\text{Vertical circular cylinder:} \qquad m_w = \rho \frac{D^3}{6} \tag{4.5}$$

Values of K_m are obtained from the graph in Figure 4.4*a*.

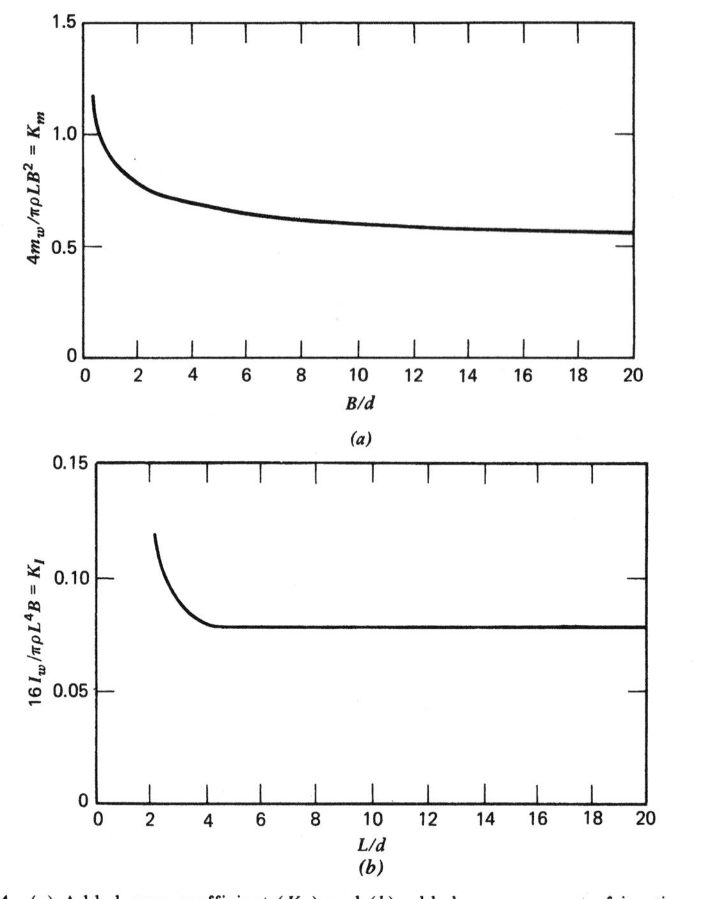

Figure 4.4 (*a*) Added-mass coefficient (K_m) and (*b*) added-mass moment of inertia coefficient (K_I) for a rectangular float.

The *natural pitching frequency* of a float is

$$f_\theta = \frac{1}{T_\theta} = \frac{\omega_\theta}{2\pi} = \frac{1}{2\pi}\sqrt{\frac{C}{I_y + I_w}} \tag{4.6}$$

where C is the *hydrostatic restoring moment coefficient,*

$$C = \rho g \int_{-L/2}^{L/2} x^2 B'(x)\,dx \tag{4.7}$$

I_y is the mass moment of inertia of the body about the rotational axis through the center of gravity, and I_w is the added-mass moment of inertia. Expressions for I_w are presented in Figures 4.2 and 4.3. For the float in Figure 4.1*a*,

$$C = \frac{\rho g B L^3}{12} \tag{4.8}$$

and

$$I_w = \frac{K_I \pi \rho L^4 B}{16} \tag{4.9}$$

where K_I values are obtained from Figure 4.4*b*, whereas for a floating *circular cylinder* with a vertical axis, the hydrostatic restoring moment and added-mass moment of inertia are, respectively,

$$C = \frac{\rho g \pi D^4}{64} \tag{4.10}$$

and

$$I_w = \frac{\rho \pi D^2 d^3}{12} \tag{4.11}$$

The importance of the frequency expressions of equations (4.3) and (4.6) is that they define the conditions for *motion resonance*. When a body with natural heaving and pitching periods described by equations (4.3) and (4.6), respectively, encounters waves that have one of these periods, a *maximum motion* in either heave or pitch can be expected. The *amplitude* of the motion will depend on the amount of *damping* in the system. In the design of a wave energy conversion device, therefore, our goal is to minimize the damping to obtain the maximum response.

Example 4.1

The bodies sketched in Figure 4.1 have a length L of 10 ft (3.05 m), a height Z of 3 ft (0.914 m), and a breadth B of 5 ft (1.52 m). The draught d of the bodies is 1.5 ft (0.457 m). The displaced water mass is

$$m = \rho L B d$$

$$= 150 \text{ lb-sec}^2/\text{ft } (2190 \text{ kg})$$

which, then, must be the mass of the body. Assuming that this mass is uniformly distributed along the length and breadth, the mass moment of inertia of the body is [see, for example, Eshbach (1975)]

$$I_y = \frac{m(L^2 + Z^2)}{12}$$

$$= 1360 \text{ lb-sec}^2\text{-ft } (1840 \text{ kg-m}^2)$$

From the results of Wendel (1956) presented in Figure 4.4*a*, the added mass of the floats is

$$m_w = \frac{0.639 \rho \pi L B^2}{4}$$

$$= 251 \text{ lb-sec}^2/\text{ft } (3650 \text{ kg})$$

and the added-mass moment of inertia is

$$I_w = \frac{0.0752 \rho \pi L^4 B}{16}$$

$$= 1480 \text{ lb-sec}^2\text{-ft } (2010 \text{ kg-m}^2)$$

(a) Since the waterplane area is

$$A_{wp} = LB$$

$$= 50 \text{ ft}^2 \ (15.2 \text{ m}^2)$$

the natural heaving period obtained from equation (4.3) is

$$T_z = 2\pi \sqrt{\frac{m + m_w}{\rho g A_{wp}}}$$

$$= 2.217^* \text{ sec}$$

In deep water the wavelength of the resonant wave is determined from equation (2.6) to be

$$\lambda = \frac{g T_z^2}{2\pi}$$

$$= 25.19^* \text{ ft } (7.70 \text{ m}) = 2.519L$$

Thus the wave approximately satisfies the optimal condition of equation (4.2).

(b) From the results of equation (4.8), the restoring moment for the floats in Figure 4.1 is

$$C = \frac{\rho g L^3 B}{12}$$

$$= 26{,}800 \text{ lb-ft } (36{,}300 \text{ N-m})$$

From equation (4.6) the natural pitching period is

$$T_\theta = 2\pi \sqrt{\frac{I_y + I_w}{C}}$$

$$= 2.04 \text{ sec}$$

*The accuracy is spurious; however, this accuracy is required in the energy calculations of Example 4.3.

and the wavelength of the deep water resonant wave is

$$\lambda = \frac{gT_\theta^2}{2\pi}$$

$$= 21.3 \text{ ft } (6.50 \text{ m})$$

Some wave energy conversion floats are designed to ride on the surface of large waves or swells. The design principle in the case of these *wave riders* is simply to have the natural heaving and pitching frequencies much higher than the wave frequency. Referring to equations (4.3) and (4.6), respectively, it can be seen that mass and mass moment of inertia of the float in question must both be relatively small, whereas the restoring force and moment values are large.

Example 4.2

The heaving body in Example 4.1 has a natural heaving period of 2.217 sec. The resonant wave has a length of 25.2 ft (7.70 m). This body in a 10-sec wave would act as a surface rider since, from equation (2.6), the wavelength is

$$\lambda = \frac{gT^2}{2\pi}$$

$$= 512 \text{ ft } (156 \text{ m})$$

The length of the float in Example 4.1 is 10 ft (3.05 m); thus $L/\lambda = 0.0195$. A secondary condition for a wave rider is, then, that $L/\lambda \ll 1$.

The *equations of motion* can be used to predict the heaving and pitching responses of floating bodies (i.e., Z and θ in Figure 4.1). These equations are somewhat complex, and the reader is referred to the text by McCormick (1973) for their derivations. In general, we can represent the heaving displacement of a body by

$$z = \frac{(F_o/\rho g A_{wp})\cos(\omega t + \gamma - \sigma_z)}{\sqrt{(1 - \omega^2/\omega_z^2)^2 + (2\Delta_z \omega/\omega_z)^2}}$$

$$= Z_0 \cos(\omega t + \gamma - \sigma_z) \qquad (4.12)$$

where F_0 is the *wave force amplitude* [see equation (4.28)], Z_0 is the *motion amplitude*, ω is the circular wave frequency, t is time in seconds, γ is a phase angle that depends on the wave force components, and σ is a phase angle that depends primarily on the dimensionless system *damping factor* Δ_z. This damp-

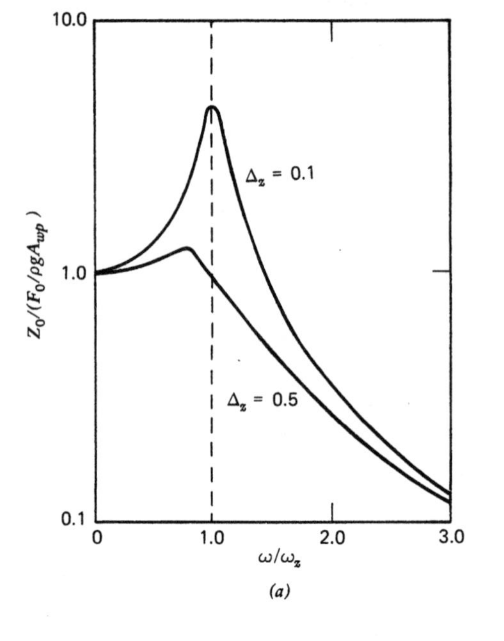

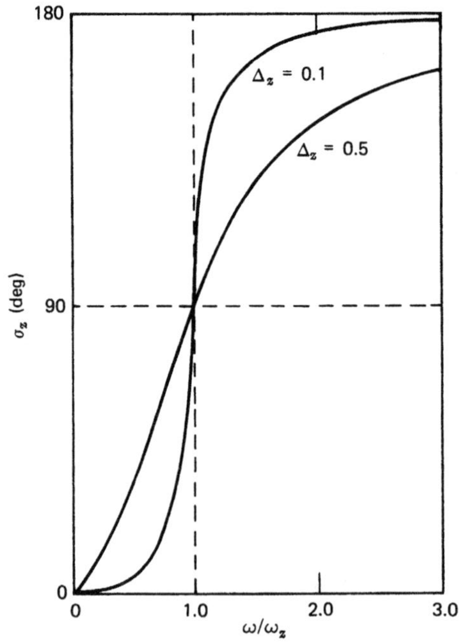

Figure 4.5 Magnification factor and phase angle variation with frequency ratio for two damping ratio values: (*a*) magnification factor as a function of frequency ratio; (*b*) phase angle of equation (4.12) as a function of frequency ratio. Although the functions presented in this figure are for heaving, the pitching functions behave in a similar manner. From McCormick (1973).

ing factor represents the effects of both viscosity and damping waves created by the body motion itself. Plots of the *magnification factor* $Z_0/(F_0/\rho g A_{wp})$ and the phase angle σ_z are presented in Figure 4.5 as functions of the frequency ratio ω/ω_z. These plots are taken from the book by McCormick (1973).

***Note*:** In Figure 4.5*b*, $\sigma_z = 90°$ when *resonance* occurs, i.e., when $\omega/\omega_Z = 1$. Furthermore, the phase angle γ is always equal to zero when the body is *symmetric* about the x–z and y–z planes. These two conditions are assumed throughout the remainder of this section.

The *velocity* and the *acceleration* of the heaving body are obtained from

$$v_z = \frac{dz}{dt} = -\omega Z_0 \sin(\omega t + \gamma - \sigma_z) \tag{4.13}$$

and

$$\begin{aligned} \frac{d^2z}{dt^2} &= -\omega^2 Z_0 \cos(\omega t + \gamma - \sigma_z) \\ &= -\omega^2 z \end{aligned} \tag{4.14}$$

respectively.

The *kinetic energy* of the mass of the heaving system (the mass of the body m plus the added mass m_w) is

$$\begin{aligned} E_{KZ} &= \tfrac{1}{2}(m + m_w)\left(\frac{dz}{dt}\right)^2 \\ &= \tfrac{1}{2}(m + m_w)\omega^2 Z_0^2 \sin^2(\omega t + \gamma - \sigma_z) \end{aligned} \tag{4.15}$$

and the *potential energy* is

$$\begin{aligned} E_{PZ} &= \tfrac{1}{2}\rho g A_{wp} z^2 \\ &= \tfrac{1}{2}\rho g A_{wp} Z_0^2 \cos^2(\omega t + \gamma - \sigma_z) \end{aligned} \tag{4.16}$$

The *total energy* of the heaving system is then

$$\begin{aligned} E_Z &= E_{KZ} + E_{PZ} \\ &= \tfrac{1}{2}\left[(m + m_w)\omega^2 + \rho g A_{wp}\right] Z_0^2 \end{aligned} \tag{4.17}$$

Similarly, from the results of McCormick (1973) the *angular displacement*, the *velocity* and the *acceleration*, of a *purely pitching body* are, respectively,

$$\theta = \frac{(M_0/C)\cos(\omega t + \delta - \sigma_\theta)}{\sqrt{\left(1 - \omega^2/\omega_\theta^2\right)^2 + (2\Delta_\theta \omega/\omega_\theta)^2}}$$

$$= \theta_0 \cos(\omega t + \delta - \sigma_\theta) \tag{4.18}$$

$$\frac{d\theta}{dt} = -\omega\theta_0 \sin(\omega t + \delta - \sigma_\theta) \tag{4.19}$$

$$\frac{d^2\theta}{dt^2} = -\omega^2\theta_0 \cos(\omega t + \delta - \sigma_\theta)$$

$$= -\omega^2\theta \tag{4.20}$$

where θ_0 is the *pitching amplitude* and δ is the phase angle that depends on the components of the wave-induced moment. In equation (4.18) M_0 is the *moment amplitude* of equation (4.32) and the coefficient

$$C = \frac{gI_y}{d} \tag{4.21}$$

is the restoring moment, assuming the draft d to be constant (a *flat bottomed* float), ω_θ is the *natural pitching frequency* of equation (4.6), and Δ_θ is the *dimensionless damping factor.* Plots of the *magnification factor* $\theta_0/(M_0/C)$ and the phase angle σ_θ as functions of the frequency ratio ω/ω_θ are similar to those for heaving shown in Figure 4.5.

Note: From the results in Figure 4.5*b* we see that $\sigma_\theta = 90°$ at *resonance*. Furthermore, for the case of x–z and y–z body *symmetry* $\delta = -90°$. Again, in the examples these two conditions are assumed to exist.

The *kinetic energy* of a pitching body is

$$E_{K\theta} = \tfrac{1}{2}(I_y + I_w)\left(\frac{d\theta}{dt}\right)^2$$

$$= \tfrac{1}{2}(I_y + I_w)\omega^2\theta_0^2 \sin^2(\omega t + \delta - \sigma_\theta) \tag{4.22}$$

and the *potential energy* is

$$E_{P\theta} = \tfrac{1}{2}C\theta^2$$

$$= \tfrac{1}{2}C\theta_0^2 \cos^2(\omega t + \delta - \sigma_\theta) \tag{4.23}$$

where equation (4.21) can be used to replace the restoring moment coefficient

C if the float is flat bottomed. The total pitching energy is

$$E_\theta = E_{K\theta} + E_{P\theta}$$
$$= \tfrac{1}{2}\left[(I_y + I_w)\omega^2 + C\right]\theta_0^2 \qquad (4.24)$$

The *mechanical power* available from a body in pure heave is the product of the wave-induced force F_z and the heaving velocity of equation (4.13). To determine the wave-induced force, we must first define the geometry of the float. We confine our attention to *flat-bottomed floats* that have either *rectangular* or *circular waterplane areas*. Using the "strip theory" described in McCormick (1973) and elsewhere, the wave-induced force on a *rectangular float* in a *linear wave* is

$$F_{zR} = \frac{\rho g H B \lambda}{2\pi}\left(e^{-2\pi d/\lambda} + 1\right)\sin\left(\frac{\pi L}{\lambda}\right)\cos(\omega t) \qquad (4.25)$$

referring to Figure 4.1 for notation. For a heaving *circular float* the force is approximated by

$$F_{zc} = \frac{\rho g H \pi R^2}{4}\left(1 - \frac{\pi^2 R^2}{2\lambda^2}\right)\left(e^{-2\pi d/\lambda} + 1\right)\cos(\omega t) \qquad (4.26)$$

where R is the waterplane radius; that is, the waterplane area is

$$A_{wp} = \pi R^2 \qquad (4.27)$$

The force expressions in both equations (4.25) and (4.26) can be represented by

$$F_z = F_0 \cos(\omega t) \qquad (4.28)$$

where F_0 is the *force amplitude* in equation (4.12). The *power* of a purely heaving *symmetric* body in *resonance* with the wave is then

$$P_z = F_z \frac{dz}{dt} \qquad (4.29a)$$

and the *average power* over one wave period is

$$\hat{P}_z = \frac{1}{T}\int_0^T P_z\, dt$$
$$= \frac{F_0 \omega Z_0}{2} \qquad (4.29b)$$

where $\sigma_z = 90°$ in equation (4.12) as a result of the resonance and $\gamma = 0$ because of the symmetry with respect to the x–z and y–z planes. Dimension-

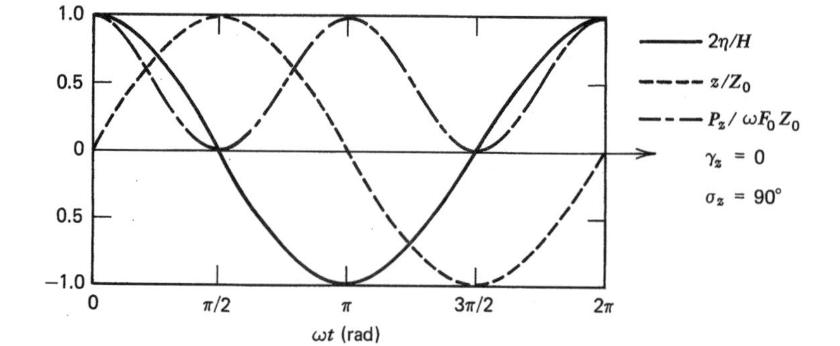

Figure 4.6 Dimensionless power, heaving displacement, and free-surface displacement for a body in a sinusoidal under a condition of resonance.

less plots of the power, body displacement, and free-surface displacement are presented in Figure 4.6 using the expressions of equations (4.29a), (4.13), and (2.1), respectively.

The *wave-induced moment* on the body in Figure 4.1*b* depends on the shape of the waterplane (assuming the draft d to be constant). For the *rectangular waterplane*

$$M_{\theta R}\circlearrowright^{+} = \frac{\rho g H B \lambda}{4\pi}\left(e^{-2\pi d/\lambda} + 1\right)\left[\frac{\lambda}{\pi}\sin\left(\frac{\pi L}{\lambda}\right) - L\cos\left(\frac{\pi L}{\lambda}\right)\right]\sin(\omega t) \quad (4.30)$$

whereas the moment on the *circular cylinder* is approximately

$$M_{\theta C}\circlearrowright^{+} = \rho g H \pi^2 R^2\left(e^{-2\pi d/\lambda} + 1\right)\sin(\omega t) \quad (4.31)$$

The moment expressions in equations (4.30) and (4.31) can be written as

$$M_{\theta}\circlearrowright^{+} = M_0 \sin(\omega t) \quad (4.32)$$

The *mechanical power* available from a *purely pitching symmetric body* is the product of the moment of equation (4.32) and the angular velocity of equation (4.19):

$$P_\theta = M_\theta \frac{d\theta}{dt}$$

$$= M_0 \omega \theta_0 \sin^2(\omega t) \quad (4.33)$$

Example 4.3

The heaving body in Examples 4.1 and 4.2 is in *resonance* with the 2.217-sec wave and has a heaving amplitude Z_0 of 2 ft (0.610 m). Thus the phase angle σ_z in equations (4.12) through (4.14) is 90° because of the resonance condition. The body is assumed to be symmetric about the

x–z and y–z planes, so that $\gamma = 0$ in these same equations. The *mechanical energy* of the heaving body is obtained from equation (4.17):

$$E_Z = \tfrac{1}{2}\left[(m + m_w)\omega^2 + \rho g A_{wp}\right] Z_0^2$$

$$= \tfrac{1}{2}\left[(150 + 251)\left(\frac{2\pi}{2.217}\right)^2 + 2.00(32.2)50\right]2^2$$

$$= 6440 \text{ lb-ft } (8760 \text{ N-m})$$

The total *energy of the wave* of height $H = 4$ ft (0.610 m) and crest length b (equal to the width of the body width B) is obtained from equation (2.17):

$$E = \frac{\rho g H^2 \lambda B}{8}$$

$$= \frac{2.00(32.2)4^2(25.19)5}{8}$$

$$= 16{,}220 \text{ lb-ft } (22{,}060 \text{ N-m})$$

Thus the *efficiency* of the heaving body is

$$\epsilon_z = \frac{E_z}{E}$$

$$= \frac{6440}{16{,}220}$$

$$= 39.7\% \tag{4.34}$$

The *average mechanical power* of the heaving motions over one wave period is obtained from equation (4.29b).

$$\hat{P}_z = \frac{F_0 \omega Z_0}{2}$$

$$= \left[\frac{\rho g H B \lambda}{2\pi}\left(e^{-2\pi d/\lambda} + 1\right)\sin\left(\frac{\pi L}{\lambda}\right)\right]\frac{\omega Z_0}{2}$$

$$= \left[\frac{2.00(32.2)4(5)25.19}{2\pi}\left(e^{-2\pi(1.5)/25.19} + 1\right)\sin\left(\frac{\pi 10}{25.19}\right)\right]\frac{2.83(2)}{2}$$

$$= 23{,}390 \text{ lb-ft/sec } (31.7 \text{ kW})$$

Heaving and pitching bodies are the most common wave energy conversion devices. One patented device is sketched in Figure 4.7, and several

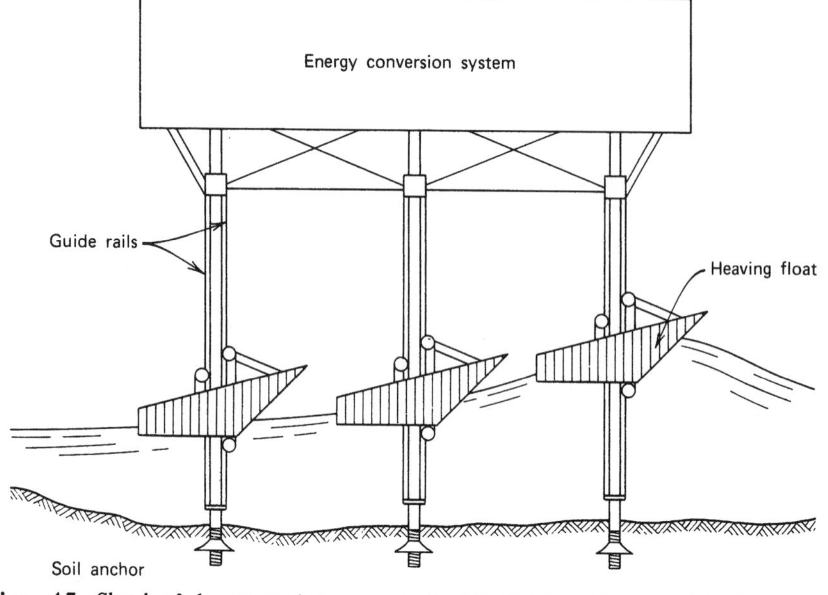

Figure 4.7 Sketch of the patented "wave motor" of P. Wright, U.S. Patent Number 599,756, March 1, 1898. The purely heaving configuration.

variations are sketched in Figure 4.8. Although these bodies are efficient when in resonance with a monochromatic wave, they are less efficient in off-design waves and in random seas. The ease of mooring (usually a single-point mooring) and the simplistic design make these bodies most attractive from a cost standpoint. Furthermore, the *wave focusing* aspect, discussed in Section 4.2, D, which occurs under a resonance condition, adds to the attractiveness of these devices.

The *energy conversion subsystems* that can be used in conjunction with heaving and pitching bodies include linear inductance devices, protonic devices, piezoelectric devices, compressed air/accumulator systems, and rotational inductance systems. These subsystems are described and discussed in Chapter 5.

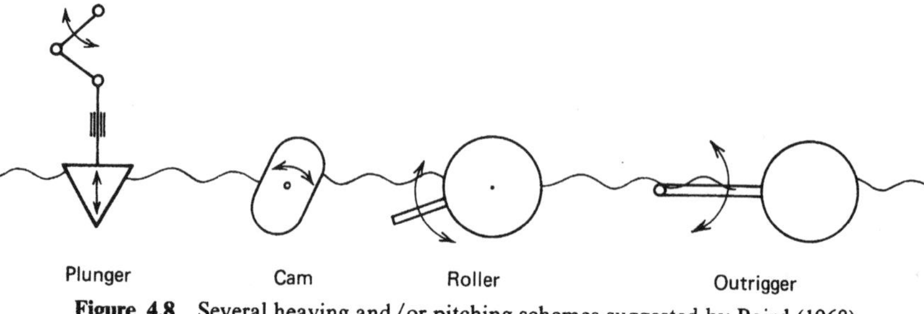

Figure 4.8 Several heaving and/or pitching schemes suggested by Baird (1968).

B Cavity Resonators

A second type of resonance that can be used in the conversion of ocean wave energy is *cavity resonance.* The most contempory use of this phenomenon in wave energy conversion is due to efforts of Masuda (1971) of the Japan Marine Science and Technology Center and R. M. Ricafranca of RMR Research and Engineering Services in the Philippines. These two investigators, working independently, are responsible for the first commercially available wave energy conversion systems designed to provide power to navigation aids such as light buoys. Furthermore, primarily due to Masuda's efforts, cavity resonance is used in the first full-scale wave energy project conducted jointly by Canada, Ireland, Japan, the United Kingdom, and the United States under the auspices of the International Energy Agency (IEA).

McCormick (1974a, 1974b, 1976) theoretically studied cavity resonance with monochromatic waves and with Carson and Rau (1975), performed experimental studies of the phenomenon at the U.S. Naval Academy. More recently Masuda et al. (1978) examined the pneumatic or cavity resonance system in preparation for the IEA study.

To gain an understanding of the use of cavity resonance in wave energy conversion, consider the vertical circular pipe shown in Figure 4.9. The pipe, which is either fixed in position or passes through the center of a floatation collar, is capped at the top and a small orifice is positioned in the center of the cap. If the pipe is fixed in position the water column in the center will *resonate* with a wave having a *frequency* of

$$f_c = \frac{\omega_c}{2\pi} = \frac{1}{T_c} = \frac{1}{2\pi}\sqrt{\frac{g}{L_1 + L_1'}} \tag{4.35}$$

where, referring to Figure 4.9, L_1 is the still water length of the water column and L_1' is an "effective" length due to the added mass excited by the water column. When resonance occurs with a water wave, the water column oscillates with an amplitude of $H_1/2$ (assuming *sinusoidal motions*), where the motion double amplitude H_1 is normally greater than the height H of the exciting wave. The water column thus acts as a *piston* and causes an oscillatory motion of the air column above the internal free surface.

From the results of McCormick (1974a) we assume that the water column has an average *vertical displacement* of

$$z_1 = \frac{\overline{H}_1}{2}\cos(\omega t) \tag{4.36}$$

Thus the *velocity* and ths *acceleration* are, respectively,

$$v_1 = \frac{dz_1}{dt} = -\frac{\omega \overline{H}_1}{2}\sin(\omega t) \tag{4.37}$$

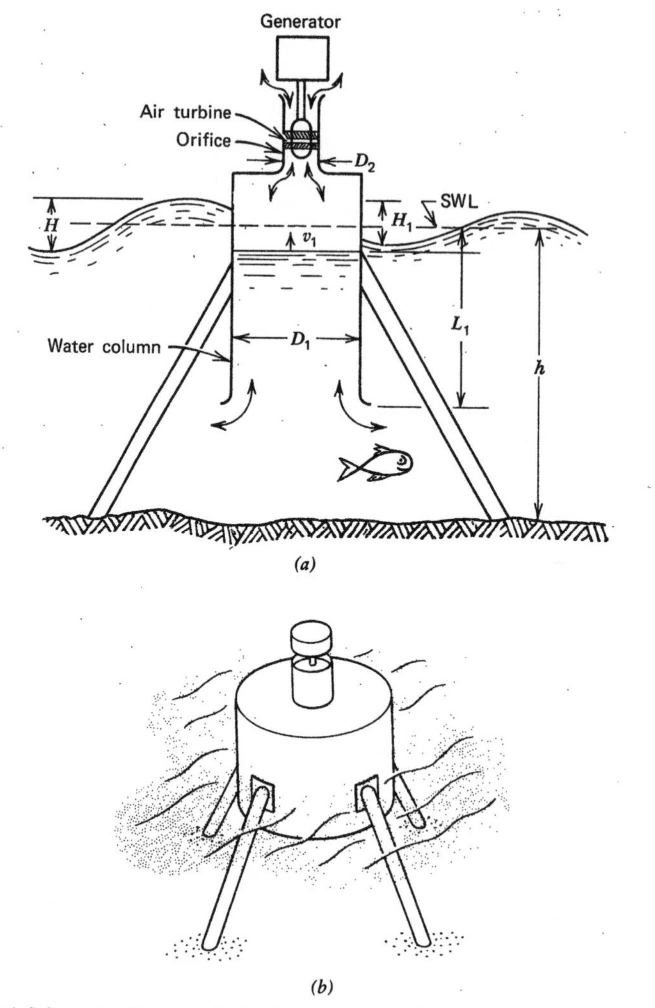

Figure 4.9 (*a*) Schematic diagram of a stationary pneumatic wave energy convertor; (*b*) sketch of a stationary pneumatic wave energy convertor.

and

$$\frac{d^2 z_1}{dt^2} = -\omega^2 z_1 \tag{4.38}$$

The air just above the water column will also have vertical motions described by equations (4.36), (4.37), and (4.38). The air adjacent to the cap is forced through the *orifice* of area A_2. Assuming the airflow to be *incompressible*, the air velocity in the orifice (i.e., v_2) is obtained from the mathematical expression of the conversion of mass called the *equation of continuity*.

$$v_2 = v_1 \frac{A_1}{A_2} \tag{4.39}$$

where A_1 is the cross-sectional area of the water column, that is,

$$A_1 = \frac{\pi D_1^2}{4} \tag{4.40}$$

Since the orifice area A_2 is much smaller than the water column area A_1, the orifice air velocity v_2 is much greater than the average vertical velocity of the air above the water column. This high-orifice velocity can be exploited by placing a *double-acting turbine* in the orifice to convert the kinetic energy of the air into electrical energy. This, then, is the principle of *cavity resonance wave energy conversion*.

Example 4.4

The vertical cylinder in Figure 4.9 has a diameter D_1 of 3 ft (0.914 m) and an orifice diameter D_2 of 0.5 ft (0.152 m). The length of the resting water column L_1 is 10 ft (3.05 m). The water column is in resonance with the incident wave and has a double amplitude of motion H_1 of 2 ft (0.610 m). From equation (4.35), assuming L_1' to be zero for the purpose of illustration, the natural frequency of the internal water motion is

$$f_c = \frac{1}{2\pi}\sqrt{\frac{g}{L_1}}$$

$$= \frac{1}{2\pi}\sqrt{\frac{32.2}{10}}$$

$$= 0.286 \text{ Hz}$$

and

$$\omega_c = 2\pi f_c$$

$$= 1.80 \text{ rad/sec}$$

The velocity of the water column is obtained from equation (4.37).

$$v_1 = -\frac{\omega_{c_2} H_1}{2}\sin(\omega_c t)$$

$$= -\frac{1.80(2)}{2}\sin(1.80t)$$

$$= -1.80\sin(1.80t) \text{ ft/sec}$$

or

$$v_1 = -0.549 \sin(1.80t) \text{ m/sec}$$

Note: The negative sign simply indicates the phase relation with the displacement. The air velocity within the orifice is obtained from equation (4.39).

$$v_2 = v_1 \frac{A_1}{A_2}$$

$$= v_1 \frac{(\pi D_1^2/4)}{(\pi D_2^2/4)} = v_1 \frac{D_1^2}{D_2^2}$$

$$= -1.80(9/0.25)\sin(1.80t)$$

$$= -64.8 \sin(1.80t) \text{ ft/sec}$$

$$= -19.8 \sin(1.80t) \text{ m/sec} \quad (4.41)$$

Thus the velocity V_2 is 36 times greater than V_1.

McCormick (1974b) derives the wave power conversion expression for a stationary wave energy convertor. Both the derivation and the power expression are too complex for presentation here. However, results obtained from the power equation are presented in Example 4.5.

Example 4.5

Consider the stationary pneumatic wave energy convertor of Figure 4.9 to have the following dimensions:

$$L_1 = 5 \text{ ft } (1.52 \text{ m})$$

$$D_1 = 40 \text{ ft } (12.2 \text{ m})$$

$$D_2 = 4 \text{ ft } (1.22 \text{ m})$$

These dimensions are those of the system analyzed by McCormick (1974b). From that reference, the wave power converted by the system operating under ideal conditions is shown as a function of wave period in Figure 4.10. The first two *peaks* in the power curve occurring at the lower periods correspond to conditions similar to those described by equations (4.1) and (4.2). Essentially, if the wavelength λ is smaller than the *characteristic length* of cross section of the water column, then an odd or even number of waves can occur within the water column. When the number of half-waves is *even*, the integrated power output is zero.

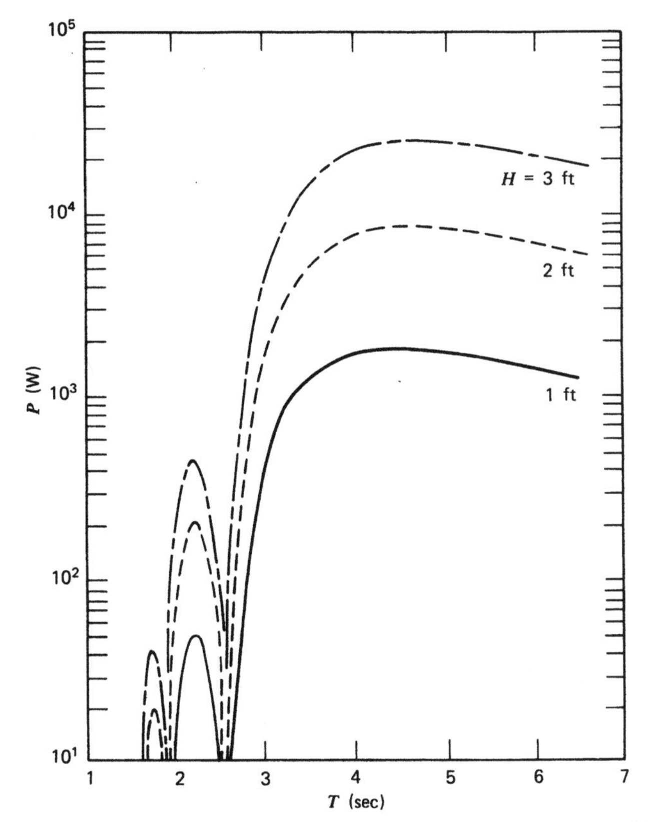

Figure 4.10 Time-averaged power versus wave period for various wave heights. From McCormick (1974b).

If the number is *odd*, a peak occurs in the power curve. When λ is greater than the characteristic length, the power is obtained from a single wave.

In Figure 4.10 we see that the maximum power conversion occurs at a period

$$T_c = 2\pi\sqrt{\frac{L_1 + L_1'}{g}}$$

$$= 4.29 \text{ sec}$$

Thus the "effective" length due to the added-mass is 1 ft (3.05 m). For a wave height of 3 ft (0.914 m), the maximum power value is 25.0 kW.

As mentioned earlier in this section, the cavity resonance technique is used in some commercially available buoys to power *navigation aids* such as lights and horns. If designed properly, this application can take advantage not only of the cavity resonance, but also of the heaving motion of the buoy. Referring to Figure 4.11, the air velocity in the orifice v_2 can be expected to have two relative maxima, one at the cavity resonance frequency of equation (4.35) and the other at the heaving resonance frequency of equation (4.3). The relationship between the air and body velocities is as follows:

$$v_2 = \frac{(v_1 - v_z)A_1}{A_2} \tag{4.42a}$$

where v_z is the body velocity of equation (4.13). From McCormick's (1976) analysis, this air velocity expression can also be written as

$$v_2 = -\omega \frac{A_1}{A_2}\left[\frac{H_1}{2}\sin(\omega t - \epsilon) - Z_0 \sin(\omega t - \sigma_z)\right] \tag{4.42b}$$

where Z_0 is the amplitude of the heaving buoy, σ_z is the phase angle discussed in the Note before equation (4.13), and ϵ is the phase angle between the external wave and the internal water surface motions. The value of ϵ depends on the system damping. For an undamped system, $\epsilon = 0$. When $\omega = \omega_c$, *cavity resonance* occurs; however, $\epsilon = 90°$ from the results of McCormick (1976). Similarly, when $\omega = \omega_z$ [equation (4.3)], the wave and buoy motions are in resonance and $\sigma_z = 90°$. By using these resonance conditions, we can now see that the *design condition* is

$$\omega_c = \omega_z \tag{4.43a}$$

or, using the results of equations (4.35) and (4.3), respectively,

$$\frac{g}{L_1 + L_1'} = -\frac{\rho g A_{wp}}{m + m_w} \tag{4.43b}$$

where, for an axisymmetric float with an outer diameter D_0 and an inner diameter D_1, the waterplane area is

$$A_{wp} = \frac{\pi}{4}\left(D_0^2 - D_1^2\right) \tag{4.44}$$

McCormick (1976) derives the expression for the power delivered to a turbine located in the orifice. Although the derivation of the expression is far too complex to present here, results of the analysis are presented in Example 4.6.

Example 4.6

McCormick (1976) presents an expression for the theoretical power output of a wave energy conversion buoy of the type shown in Figure 4.11. To illustrate the power conversion of such a system, consider the

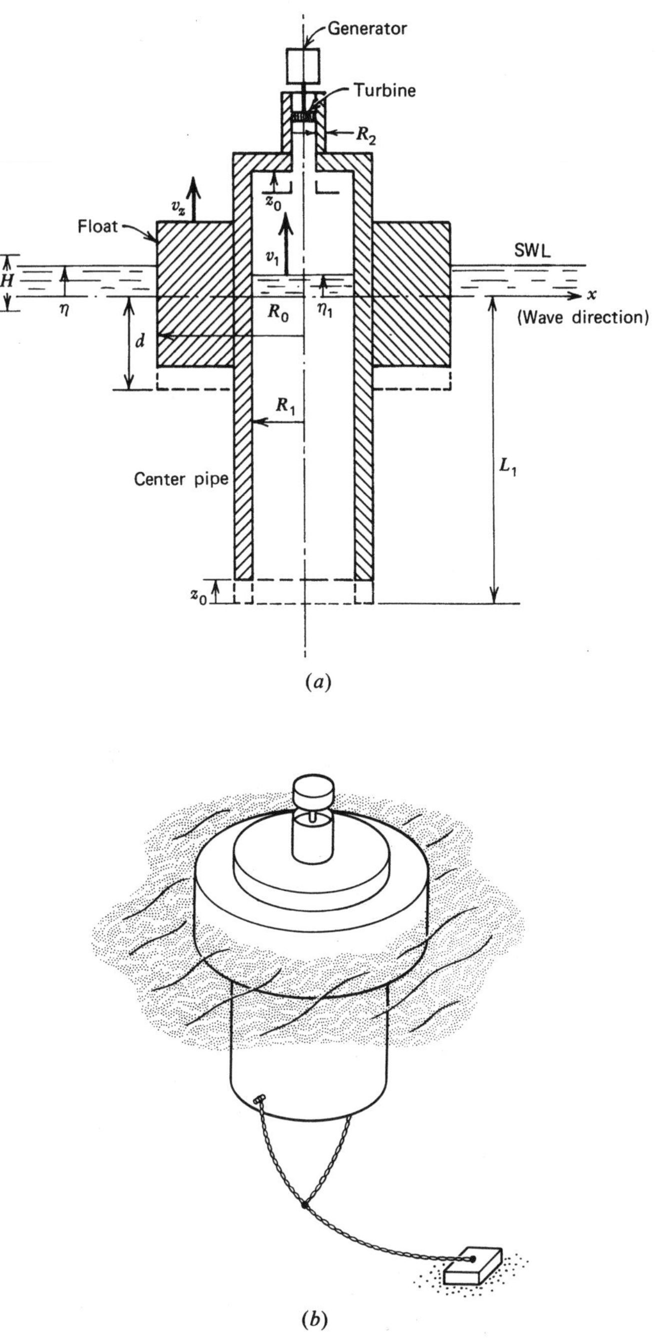

Figure 4.11 (*a*) Schematic diagram of a floating pneumatic wave energy convertor. [From McCormick (1976)]; (*b*) sketch of a floating pneumatic wave energy convertor.

buoy system to have the following properties:

$$d = 2.74 \text{ ft } (0.835 \text{ m})$$

$$D_0 = 8.00 \text{ ft } (2.44 \text{ m})$$

$$D_1 = 2.00 \text{ ft } (0.610 \text{ m})$$

$$D_2 = 0.679 \text{ ft } (0.207 \text{ m})$$

$$L_1 + L_1' = 18.0 \text{ ft } (5.49 \text{ m})$$

$$m = 258 \text{ lb-sec}^2/\text{ft } (3770 \text{ kg})$$

$$m_w = 258 \text{ lb-sec}^2/\text{ft } (3770 \text{ kg})$$

A certain amount of system damping is assumed on the basis of the experimental results of McCormick et al. (1975). The power available to the turbine in Figure 4.11 is presented in Figures 4.12 and 4.13 as a

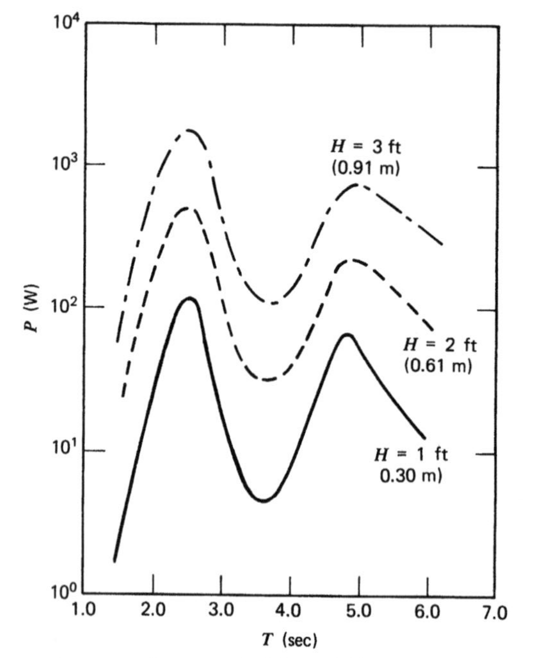

Figure 4.12 Power as a function of wave period for a 4.15-ton pneumatic wave energy conversion buoy. From McCormick (1976).

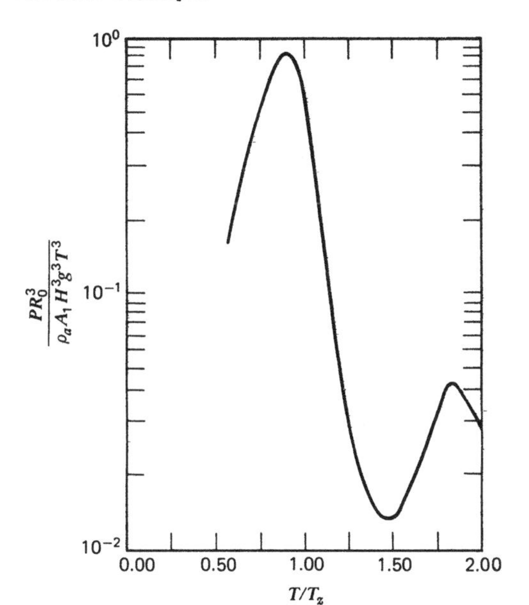

Figure 4.13 Dimensionless power curve for the prototype design of a pneumatic wave energy conversion buoy. Note: ρ_a is the mass-density of air (1.22 kg/m^3). From McCormick (1976).

function of wave period T. In Figure 4.12 the cavity resonant period is

$$T_c = 2\pi\sqrt{\frac{L_1 + L_1'}{g}}$$

$$= 2\pi\sqrt{\frac{18.0}{32.2}}$$

$$= 4.70 \text{ sec}$$

from equation (4.35), whereas the heaving resonant period obtained from equation (4.3) is

$$T_z = 2\pi\sqrt{\frac{m + m_w}{\rho g A_{wp}}}$$

$$= 2\pi\sqrt{\frac{258 + 258}{2.00(32.2)47.1}}$$

$$= 2.59 \text{ sec}$$

The results of Figure 4.12 show a peak power of approximately 1.25 kW occurring at heaving resonance and a secondary peak of 0.900 kW occurring at the cavity resonance in a 3-ft (0.914-m) wave.

Note: The peak power values are affected by the *system damping* such that as the damping increases the power decreases. For the buoy system shown in Figure 4.11, the damping is due to the creation of waves by both the heaving motion of the buoy and the vertical motion of the water column. In addition, friction on the wetted body surfaces and friction and turbulence losses in the airflow add to the damping.

To *optimize the design* of this buoy system, use the design condition given in equation (4.43b). This equation describes the condition whereby $T_c = T_z$. Since the higher-period wave has the higher power, we choose T_c to be the design period. We keep the dimensions in equation (4.43b) the same as originally used and simply change the mass m of the system by adding *ballast*. The added mass m_w is a function of geometry and will slightly change with the additional draft d. The design mass value from equation (4.43b) is

$$m = \frac{\rho g A_{wp} L_1}{g} - m_w$$

$$= \frac{2.00(32.2)47.1(18.0)}{32.2} - 258$$

$$= 1440 \text{ slugs } (21{,}000 \text{ kg})$$

To accommodate this change in mass, the draft d of the float also increases. The new, or *design draft*, is

$$d = \frac{m}{\rho A_{wp}}$$

$$= \frac{1440}{2.00(47.1)}$$

$$= 15.3 \text{ ft } (4.66 \text{ m})$$

The energy conversion efficiency of a cavity resonance system ultimately depends on the air turbine design. As mentioned in the first paragraph of Section 4.1, B, full-scale trials are being conducted in the Sea of Japan under the auspices of the IEA to determine the effectiveness of the cavity resonance systems. The turbines used in the IEA study were designed and constructed in Japan, the United Kingdom, the the United States, respectively. These designs are described and discussed in Section 5.1, B [see also Masuda and Miyazaki (1979)].

The test platform used in the IEA study is a ship-shaped barge called the *Kaimei*. This 80-m-long, 12-m-wide barge (sketched in Figure 4.14) is designed

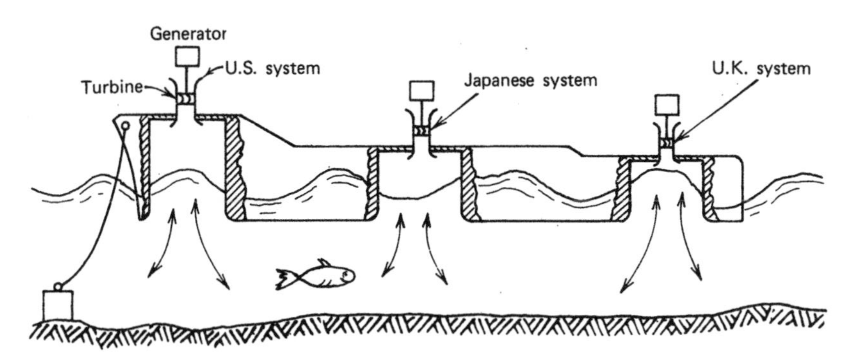

Figure 4.14 Artist's sketch of the "Kaimai" floating wave energy conversion system. From McCormick (1979).

to hold 10 turbogenerating systems and deliver a total average power of 1250 kW in the design sea. It is envisioned by the IEA that a number of Kaimei-type energy conversion systems could be moored in waters several kilometers from shore and provide the coastal communites with most of their electrical energy needs, thus permitting the wave energy conversion to relieve the national grid of some of the load.

An advance in cavity resonance wave energy conversion has recently occurred in the United Kingdom, as described by Moody (1979). Engineers have found that the available wave energy to the cavity is doubled if the cavity opening faces the oncoming wave. In this situation the wave pressure at the opening is not just the static pressure but the total wave pressure. A device designed to utilize the total pressure, called the *oscillating water column*, is illustrated in Plate 4.1.

C Pressure Devices

The pressure beneath a wave is constantly varying because of the change in water level (affecting the *hydrostatic pressure*) and the motion of the water particles (affective the *dynamic pressure*). These varying pressures can be represented by *Bernoulli's equation* (Streeter, 1971)

$$p = -\rho \frac{\partial \varphi}{\partial t} - \rho g z - \tfrac{1}{2}\rho v^2 \tag{4.45}$$

where φ is the *velocity potential*, ρ is the mass density of saltwater (2.00 lb-sec^2/ft^4 or 1030 kg/m^3), z is the vertical coordinate with its origin on the SWL, and v is the water particle velocity. From McCormick (1973) and elsewhere, the velocity potential expression for a *linear wave* is

$$\varphi = \frac{gH}{2\omega} \frac{\cosh(kz + kh)}{\cosh(kh)} \sin(kx - \omega t) \tag{4.46}$$

Plate 4.1 An artist's concept of an oscillating water column wave energy conversion system. Courtesy of the Energy Technology Support Unit, AERE Harwell—The United Kingdom.

where ω is the circular frequency of the wave, h is the water depth, and k is the wave number $(2\pi/\lambda)$. Also, for the linear wave, we assume

$$v^2 \simeq 0 \tag{4.47}$$

We confine our attention to the linear wave $(H/\lambda \ll 1)$ since our interest is in designing a system to convert the energy of the *swell*. The pressure at any point beneath the wave is then

$$\begin{aligned} p &\simeq -\rho \frac{\partial \varphi}{\partial t} - \rho g z \\ &= \frac{\rho g H}{2} \frac{\cosh(kz + kh)}{\cosh(kh)} \cos(kx - \omega t) - \rho g z \end{aligned} \tag{4.48}$$

where the results of equations (4.45), (4.46), and (4.47) have been combined.

Consider the system sketched in Figure 4.15a, where a *compliant surface* (*membrane*) responds to pressure changes caused by the passage of a wave. We refer to this system as a *uniform pressure wave pump* since the wave-induced pressure within the water chamber is approximately uniform at any time t. Assume that the water chamber is *rectangular* with a length L and a

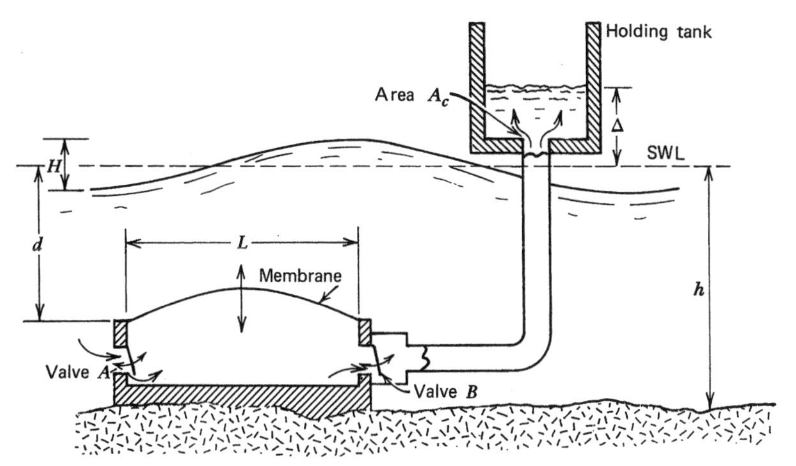

Figure 4.15a Uniform pressure wave pump with pumped storage.

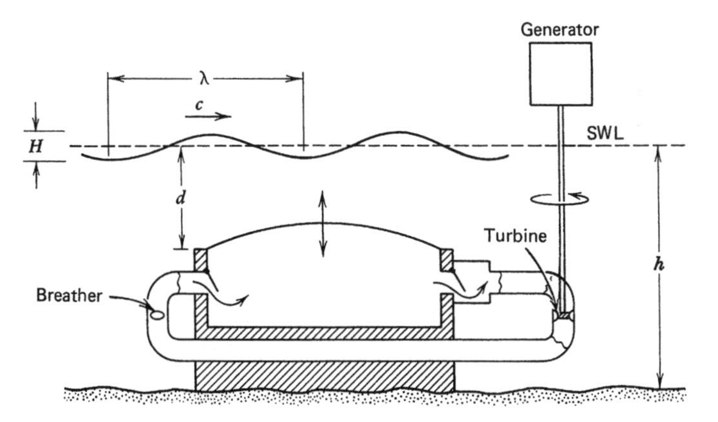

Figure 4.15b Uniform pressure wave pump with turbogenerator.

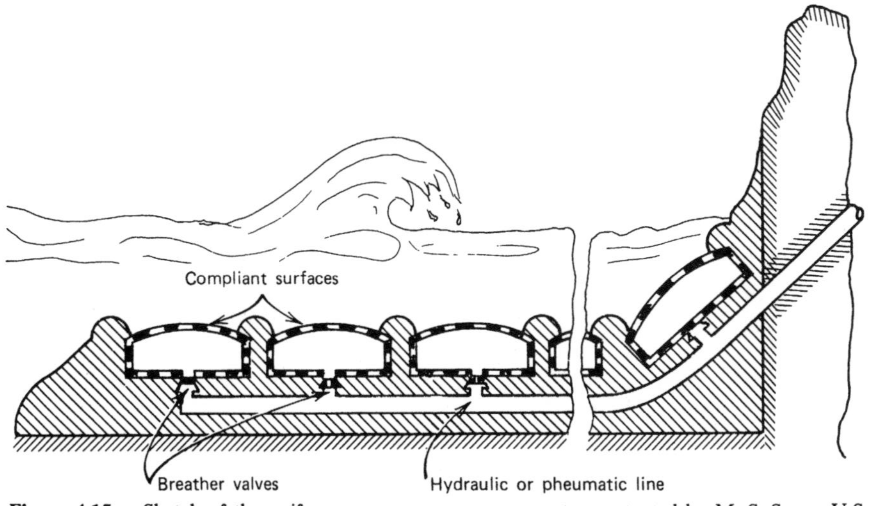

Figure 4.15c Sketch of the uniform pressure wave pump system patented by M. S. Semo, U.S. Patent Number 3,353,787, July 11, 1966.

width B. The *pressure force* on the rectangular membrane is obtained by integrating the pressure expression of equation (4.48) over the membrane area. Thus, taking our x origin at the center of the membrane, the pressure force is

$$
\begin{aligned}
F_P &= B\int_{-L/2}^{L/2} p\,dx \\
&= \frac{\rho g H B \lambda}{2\pi}\,\frac{\cosh(-kd+kh)}{\cosh(kh)}\sin\left(\frac{kL}{2}\right)\cos(\omega t) \\
&\quad + \rho g d B L
\end{aligned}
\tag{4.49}
$$

where d is the depth position of the compliant surface or membrane.

The *open system* in Figure 4.15a is designed to draw water through opening A on the upstroke of the membrane, and the opening has a one-way valve. The upstroke of the membrane occurs when the time-dependent term of equation (4.49) is negative; that is, the pressure is less than the *ambient hydrostatic pressure*, $\rho g d$. As the pressure increases, the membrane begins the downstroke, thus forcing valve A to close and valve B to open. Water is then forced from the chamber through opening B and into a hose which leads to a holding tank for later usage. A variation of the fluid conduit system is sketched in Figure 4.15b, where the water within the system always remains therein, and directly drives a water turbine. This system, referred to as a *closed system*, can be much smaller since no storage is required. A patented uniform pressure wave pump is shown in Figure 4.15c.

The pressure force expression of equation (4.49) is composed of a *dynamic force component* F_{PD} and a *hydrostatic component* F_{PH}; the former is time dependent in that equation. Focusing our attention on the *open system* of Figure 4.15a, we see that water enters opening A at the same hydrostatic pressure as that within the chamber. Thus the pumping is caused by the dynamic force component; that is,

$$
\begin{aligned}
F_{PD} &= \frac{\rho g H B \lambda}{2\pi}\,\frac{\cosh(kh-kd)}{\cosh(kh)}\sin\left(\frac{\pi L}{\lambda}\right)\cos(\omega t) \\
&= F_{PD0}\cos(\omega t)
\end{aligned}
\tag{4.50}
$$

To make the pump operate, the force of equation (4.50) must overcome the *back pressure* caused by the *head* Δ and the flow losses in the piping and valves. Assuming that at *equilibrium* the water surface motions in the holding tank will be in phase with the motions of the compliant surface (or membrane) of the pump, then

$$
F_{PD0} = \rho g \Delta A_c \tag{4.51}
$$

neglecting flow losses where A_c is the conduit area at C. Thus the equilibrium height Δ is obtained by combining equations (4.50) and (4.51).

$$\Delta = \frac{HB\lambda}{2\pi A_c} \frac{\cosh(kh - kd)}{\cosh(kh)} \sin\left(\frac{\pi L}{\lambda}\right) \tag{4.52}$$

The *energy* required to develop the head Δ is equal to the work performed by the pump W_P. Thus the pump work expression is

$$W_P = \frac{\rho g A_c \Delta^2}{2} \tag{4.53}$$

The *efficiency* of the system is then the ratio of W_P and the wave energy, E from equation (2.17) in which the crest width b is replaced by B:

$$\begin{aligned} \epsilon &= \frac{W_P}{E} \\ &= \frac{4A_c\Delta^2}{H^2B\lambda} \\ &= \frac{\lambda B}{\pi^2 A_c} \frac{\cosh^2(kh - kd)}{\cosh^2(kh)} \sin^2\left(\frac{\pi L}{\lambda}\right) \end{aligned} \tag{4.54}$$

From the results in equation (4.54), we see that the efficiency can be increased by decreasing area A_c.

Example 4.7

A pressure pump similar to that sketched in Figure 4.15*a* is located in deep water where the wavelength is $\lambda = 300$ ft (91.4 m) and the wave height is $H = 3.00$ ft (0.914 m). The period for this wave, from equation (2.2), is 7.65 sec. The compliant surface or membrane is located 10 ft (3.05 m) below the SWL and is 20 ft (6.10 m) long and 20 ft (6.10 m) wide, that is, $L = B = 20$ ft (6.10 m). For purposes of illustration, we have designed the conduit water-line area A_c in Figure 4.15*a* to be equal to the membrane area LB. We have also assumed *deep water* ($h > \lambda/2$), so that the following approximations are valid:

$$\cosh(kh - kd) \simeq \frac{e}{2}^{k(h-d)} \tag{4.55}$$

$$\cosh(kh) \simeq \frac{e}{2}^{kh} \tag{4.56}$$

Thus the *maximum head* developed by the wave pressure pump, from equation (4.52), is

$$\Delta \simeq \frac{H\lambda}{2\pi L} e^{-kd} \sin\left(\frac{\pi L}{\lambda}\right)$$

$$= \frac{(3.00)300}{2\pi(20)} e^{-2\pi(10)/300} \sin\left(\frac{\pi \cdot 20}{300}\right)$$

$$= 1.21 \text{ ft } (0.368 \text{ m})$$

and the *efficiency* from equation (4.54) is

$$\epsilon = \frac{4A_c\Delta^2}{H^2\lambda B}$$

$$= \frac{4(400)(1.21)^2}{3^2(300)20}$$

$$= 0.043 \quad \text{or} \quad 4.3\%$$

It can be seen from this result that the uniform pressure wave pump is rather inefficient in the conversion of wave energy to mechanical energy.

A second pressure device is that known as the *constant-force wave pump*, which is sketched in Figure 4.16. The dynamic pressure force F_{PD} on the upper surface of the large circular cylindrical piston is the same as that on the small piston or *plunger*. The pressure on the small piston is then amplified by the hydraulic equation

$$p_2 = p_1 \frac{A_1}{A_2}$$

$$= p_1 \frac{D_1^2}{D_2^2} \tag{4.57}$$

where A_1 and D_1 are the area and diameter of the large piston, respectively, and A_2 and D_2 those of the plunger.

The maximum head developed Δ is due to the dynamic pressure force acting on the large piston, assuming the piston to be *neutrally* buoyant. This force on the circular cylindrical piston is obtained by integrating the pressure of equation (4.48) over the piston surface. Thus

$$F_{PD} = \frac{\rho g H \pi D_1^2}{16}\left(1 - \frac{\pi D_1^2}{8\lambda^2}\right) e^{-kd} \cos(\omega t)$$

$$= F_{PD0} \cos(\omega t) \tag{4.58}$$

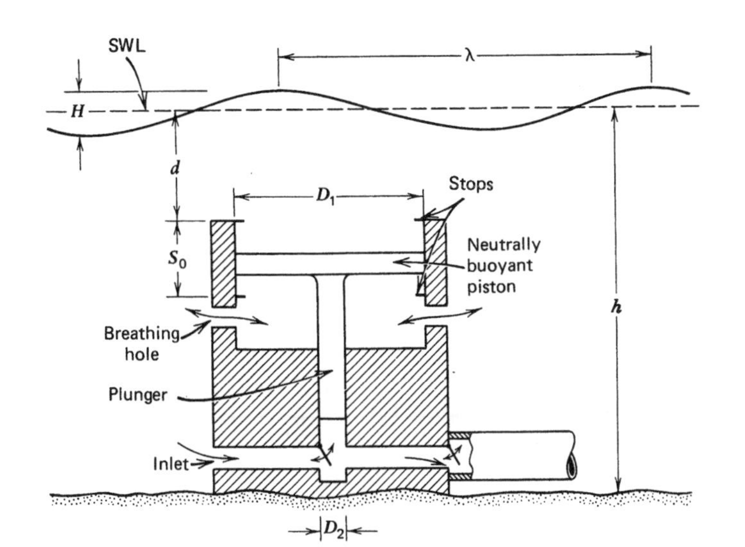

Figure 4.16*a* Schematic diagram of a constant-force wave pump.

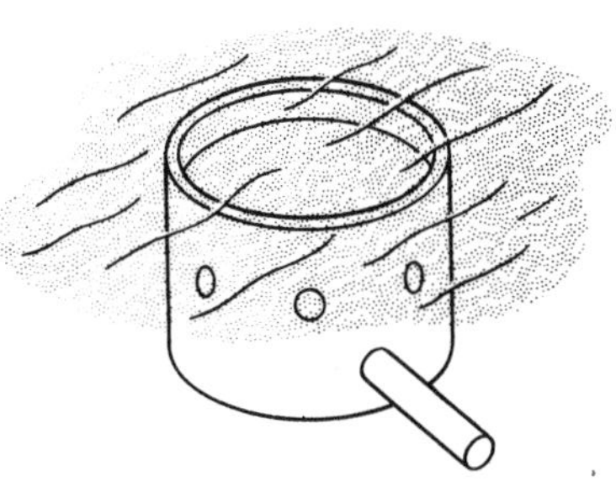

Figure 4.16*b* Sketch of a constant-force wave pump.

The *maximum head* is then obtained by combining the results of equations (4.51) and (4.58):

$$\Delta = \frac{F_{PD0}}{\rho g A_c}$$

$$= \frac{H\pi D_1^2}{16A_c}\left(1 - \frac{\pi D_1^2}{8\lambda^2}\right)e^{-kd} \tag{4.59}$$

The *efficiency* of this system is

$$\epsilon = \frac{W_P}{E}$$

$$= \frac{\rho g A_c \Delta^2}{2E}$$

$$= \frac{4A_c\Delta^2}{H^2\lambda D_1} \tag{4.60}$$

where the *pump work* W_P is obtained from equation (4.53) and the crest width b in equation (2.17) is replaced by the piston diameter D_1.

Example 4.8

For the wave conditions in Example 4.7, consider a large piston to have an area $A_1 = 400$ ft^2 (37.2 m^2), that is, $D_1 = 22.6$ ft (6.88 m). Assuming $A_c = A_1 = \pi D_1^2/4$, the maximum head developed is obtained from equation (4.59):

$$\Delta = \frac{H\pi D_1^2}{16A_c}\left(1 - \frac{\pi D_1^2}{8\lambda^2}\right)e^{-2\pi d/\lambda}$$

$$= \frac{3\pi(22.6)^2}{16(400)}\left(1 - \frac{\pi(22.6)^2}{8(300)^2}\right)e^{-2\pi(10)/300}$$

$$= 0.610 \text{ ft } (0.186 \text{ m})$$

The efficiency is obtained from equation (4.60):

$$\epsilon = \frac{4A_c\Delta^2}{H^2\lambda D_1}$$

$$= 0.0293 \qquad \text{or} \qquad 2.93\%$$

Again, the efficiency of the constant-force wave pump is not significant.

From the results in Examples 4.7 and 4.8, one can conclude that pressure devices are not very efficient. Furthermore, because of the massive construction involved in these systems the pressure devices are cost-ineffective. Thus systems similar to those sketched in Figure 4.15 and 4.16 are not strong candidates for wave energy conversion. A floating pressure device that has some promise is shown in Plate 4.2.

D Surging-Wave Energy Convertors

In Section 2.1, equation (2.14) describes the horizontal particle velocity in *shallow water*, that is, where $h/\lambda < \frac{1}{20}$.

$$u = \frac{H}{2}\sqrt{\frac{g}{h}}\cos(kx - \omega t) \tag{2.14}$$

It can be seen that the horizontal particle velocity does not vary with vertical position. This is approximately the situation when a swell nears the surf zone on a very gradual beach. When the wave *breaks*, the condition is described by

$$u\big|_{\text{crest}} = c \tag{4.61}$$

Plate 4.2 An artist's concept of a compliant bag wave energy conversion system. Courtesy of the Energy Technology Support Unit, AERE Harwell—The United Kingdom.

where c is the phase velocity of the wave in shallow water described in equation (2.9); that is,

$$c = \sqrt{gh} \tag{2.9}$$

After the condition in equation (4.61) is satisfied, the water particles on the surface approach the shoreline more rapidly than the wave itself; however, on a very gradual beach a shallow water swell will act as a *surge* over a moderate distance before plunging on the beach. There have been a number of suggested methods for capturing the energy of the wave just as it enters the *surf zone*. Let us examine the basic principles of these methods by analyzing the energy conversion system sketched in Figure 4.17*a*. A patented system is shown in Figure 4.17*b*.

The deflector is designed to absorb some of the momentum of the surge while turning the flow upward at an angle Γ to the horizontal direction. From basic fluid mechanics (e.g., Streeter, 1971), the *force on the deflector* is

$$F_d = \rho A_d u(u - V_d)[1 - \cos(\Gamma)] \tag{4.62}$$

where A_d is the vertical flow area of the surge; in other words, referring to Figure 4.17, we have

$$A_d = (h + \eta)B_d \tag{4.63}$$

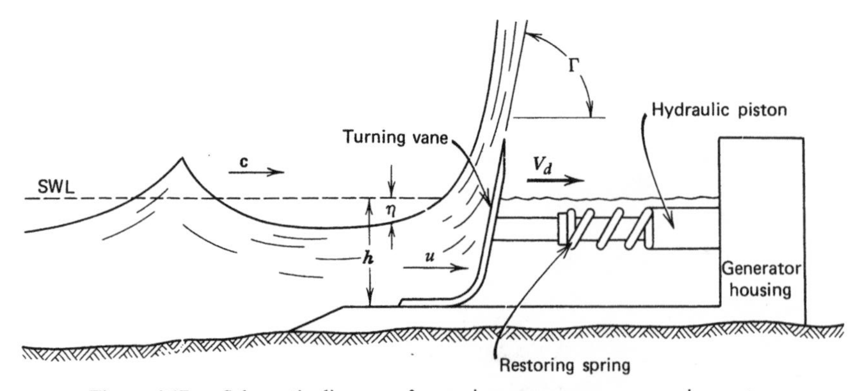

Figure 4.17*a* Schematic diagram of a surging-wave energy conversion system.

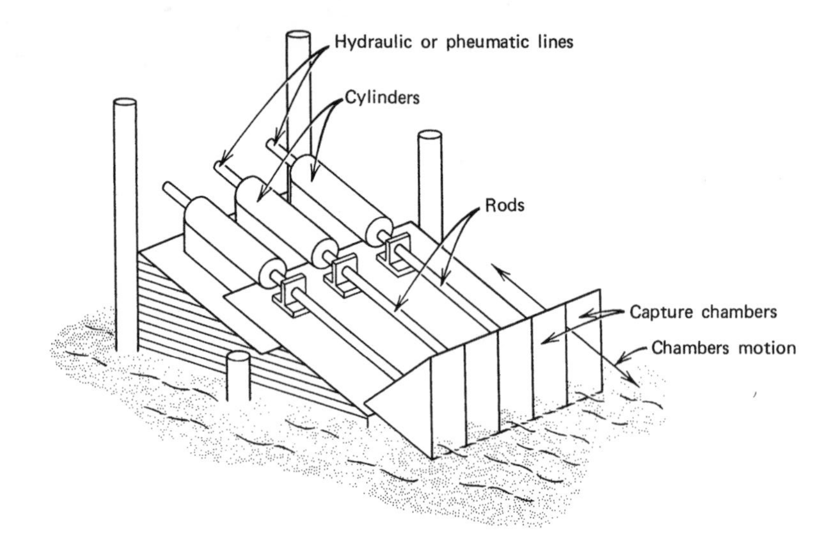

Figure 4.17*b* Sketch of a surging-wave energy conversion system patented by E. A. Wall, U.S. Patent Number 1,008,682, November 14, 1911.

where B_d is the deflector width. The displacement of the free surface η is described in equation (2.1). In equation (4.62) it can be seen that the *maximum power* occurs when $V_d = u/2$ and $\Gamma = 180°$.

Referring to the idealized *sequence* shown in Figure 4.18, we see that when the crest strikes the deflector (Figure 4.18*a*), the velocity of the deflector V_d is maximum toward the shore. The deflector then begins to decelerate until the *wave node* ($\eta = 0$) arrives, at which time the deflector stops (Figure 4.18*b*). Then the deflector accelerates seaward and attains a relative maximum velocity as the wave trough arrives (Figure 4.18*c*). Deceleration in the seaward direction then occurs until the next node arrives (Figure 4.18*d*), at which time the deflector stops and then begins its landward motion.

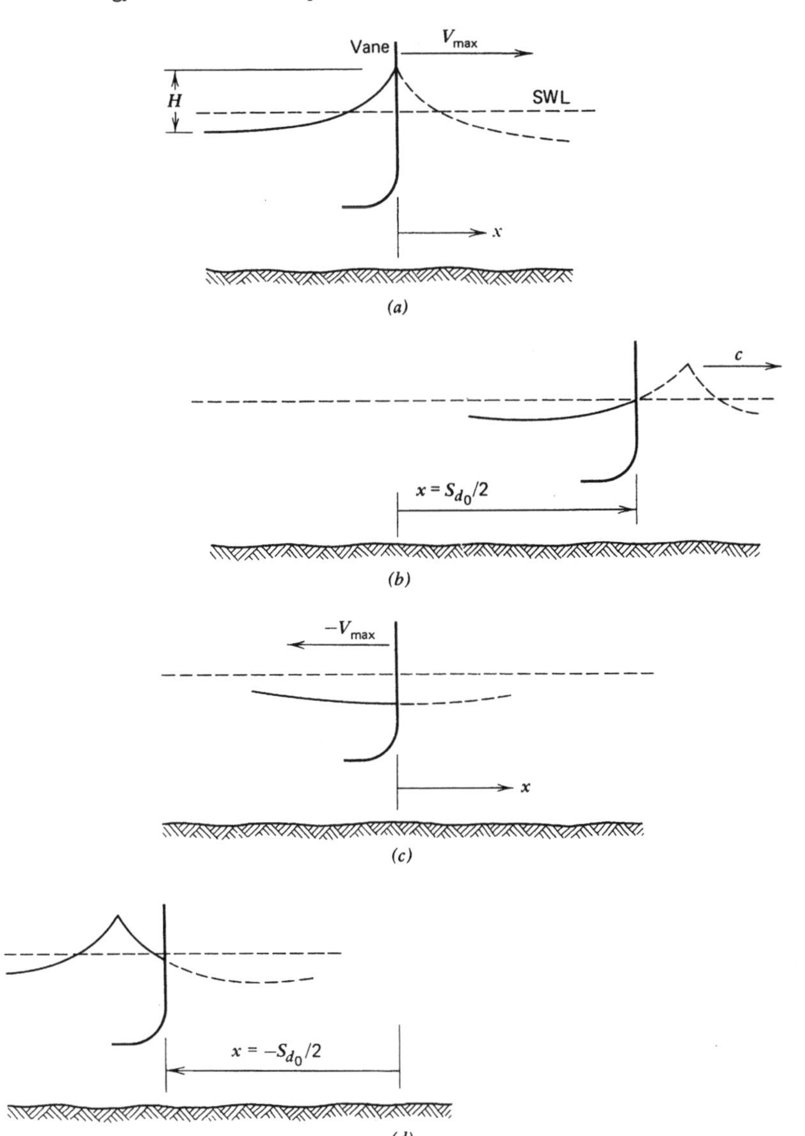

Figure 4.18 Idealistic sequence of surging-wave energy conversion device: (*a*) $\eta = H/2$ and $V_d = V_{max}$ at $x = 0$; (*b*) $\eta = 0$ and $V_d = 0$ at $x = S_{d_0}/2$; (*c*) $\eta = -H/2$ and $V_d = -V_{max}$ at $x = 0$; (*d*) $\eta = 0$ and $V_d = 0$ at $x = S_{d_0}/2$.

If the system is designed to operate within the *surf zone*, then only the situations sketched in Figures 4.18*a* and 4.18*d* are relevant since the wave behaves as a *solitary wave* of Section 2.2, B. In this case a spring or other mechanical device supplies the only restoring force to the deflector since there is no seaward motion of the water particles except for the low energy *backrush due to gravity*.

We confine our attention to waves *seaward of the break*. The wave will have a *nonlinear profile* as illustrated in Figures 2.6 and 2.7; however, for purposes of illustration, we assume the profile to be *sinusoidal* and described by equation (2.1):

$$\eta = \frac{H}{2}\cos(kx - \omega t) \tag{2.1}$$

where k is the wave number $2\pi/\lambda$ and ω is the circular wave frequency $2\pi/T$. The motions of the deflector are assumed to be *in phase* with the wave, as illustrated in Figure 4.18. Thus the *deflector velocity* is

$$V = V_0 \cos(\omega t) \tag{4.64}$$

where the deflector is located at $x = 0$ for convenience.

To obtain the expression for the force on the deflector combine equation (4.62) with equations (2.1), (2.14), (4.63), and (4.64) at $x = 0$. Thus the maximum force on the deflector ($V_d = u/2$) is obtained from

$$F_d = \rho\left[h + \frac{H}{2}\cos(\omega t)\right]B_d \cdot \frac{H}{2}\sqrt{\frac{g}{h}}\cos(\omega t)\left[\frac{H}{2}\sqrt{\frac{g}{h}} - V_0\right]\cos(\omega t) \times \left[1 - \cos(\Gamma)\right] \tag{4.65}$$

The *power* delivered to the deflector is

$$P_d = F_d V = \rho\left[h + \frac{H}{2}\cos(\omega t)\right]B_d \frac{H}{2}\sqrt{\frac{g}{h}}\, V_0 \times \left[\frac{H}{2}\sqrt{\frac{g}{h}} - V_0\right]\left[1 - \cos(\Gamma)\right]\cos^3(\omega t) \tag{4.66}$$

Example 4.9

Consider a deflector located in 10 ft (3.05 m) of water and having a breadth of 20 ft (6.10 m) and a deflector angle of 90°. The shallow water wave experienced by the deflector is assumed to have a height of 3 ft (0.914 m) and a period of 7 sec. Thus, using the results of equation (2.8), the wavelength is 125 ft (38.8 m). The *maximum force and power* occur when $\cos(\omega t) = 1$. Thus the maximum force from equation (4.65), assuming $V_0 = u_{max}/2$, is

$$F_{d_{max}} = \rho\left[h + \frac{H}{2}\right]\frac{H^2 g B_d}{8h}$$

$$= 2.00[10 + 1.5]\frac{3^2(32.2)20}{80}$$

$$= 1670 \text{ lb } (7410 \text{ N})$$

The maximum power, obtained from equation (4.66), is

$$P_{d_{max}} = F_{d_{max}} V_0 = F_{d_{max}} \frac{H}{4}\sqrt{\frac{g}{h}}$$

$$= 1670 \frac{(3)}{4}\sqrt{\frac{32.2}{10}}$$

$$= 2250 \text{ lb-ft/sec } (3.04 \text{ kW})$$

It should be noted here that the *average force* and *power* values are much less than the maximum values.

The power available from the shallow water wave is obtained from equation (2.19), using the results of equations (2.20c) and (2.9); thus

$$P = \frac{\rho g H^2 \sqrt{gh}\, B_d}{8}$$

$$= \frac{2.00(32.2)3^2\sqrt{32.2(10)}\; 20}{8}$$

$$= 26{,}000 \text{ lb-ft/sec } (35.3 \text{ kW}) \tag{4.67}$$

Comparing this value to that of the maximum power output of the device, it can be seen that the wave power conversion efficiency of the idealized device is less than 8%.

The stroke of the deflector is obtained from the *displacement* relationship

$$s_d = \int_0^t V\, dt$$

$$= V_0 \int_0^t \cos(\omega t)\, dt$$

$$= \frac{V_0}{\omega}\sin(\omega t)$$

$$= \frac{S_{d0}}{2}\sin(\omega t) \tag{4.68}$$

where S_{d0} is the *stroke*. In this example

$$S_{d0} = \frac{2V_0}{\omega}$$

$$= 2V_0 \frac{T}{2\pi}$$

$$= \frac{H}{4}\sqrt{\frac{g}{h}}\,\frac{T}{\pi}$$

$$= \frac{3}{4}\sqrt{\frac{32.2}{10}}\,\frac{7}{\pi}$$

$$= 3.00 \text{ ft } (0.914 \text{ m})$$

assuming that $V_0 = u/2$ in this idealized situation.

From the results in Example 4.7, it can be seen that the surging device operating in or near the surf zone is rather inefficient. In view of the high capital costs of such a device, ths surging wave energy convertor is not a cost-effective system.

E Particle Motion Convertors

From the discussion of wave properties in Section 2.1, we know that the water particles within *deep water* waves travel in nearly *circular orbits* with radii that decrease exponentially with depth; however, in *shallow water* the particle paths are *elliptic* with constant major axes and minor axes that decrease linearly with depth. These are illustrated in Figure 4.19, where it can be noted that the wavelength in shallow water λ is shorter than that in deep water λ_0, and that the shallow water wave height H is larger than the deep water wave height H_0.

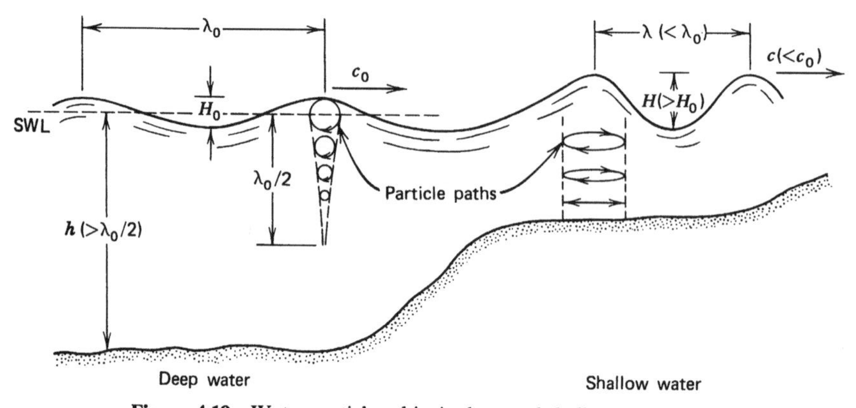

Figure 4.19 Water particle orbits in deep and shallow water waves.

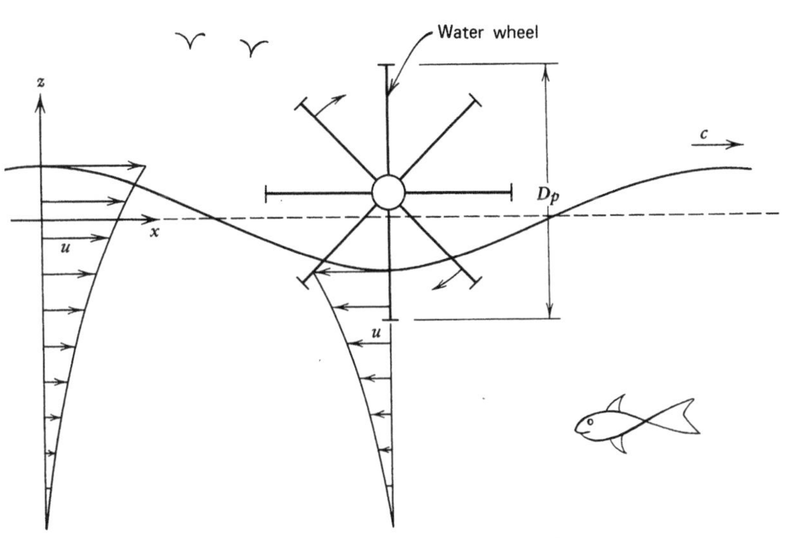

Figure 4.20 Schematic diagram of a water wheel wave energy convertor in a deep water wave.

To convert the energy of the moving water particles, we must design a device that will have motions that are approximately those of the particles. A very popular idea for such a device is the *water wheel*, which is sketched in Figures 4.20 and 4.21. The optimum design for this system operating in deep water is that for which the axis of rotation is just above the crest of the wave. In this case wave power is converted only when a *crest* passes, or over one-half of the wave period. At first glance this may not appear to be optimum to the reader since one naturally wants to convert wave power over the entire wave period. Consider the case when the wheel is fully submerged with the axis of rotation at $z = -H$. The horizontal gradient of the velocity, illustrated in Figure 4.20, is not significant over the diameter of the wheel D_p, assuming D_p

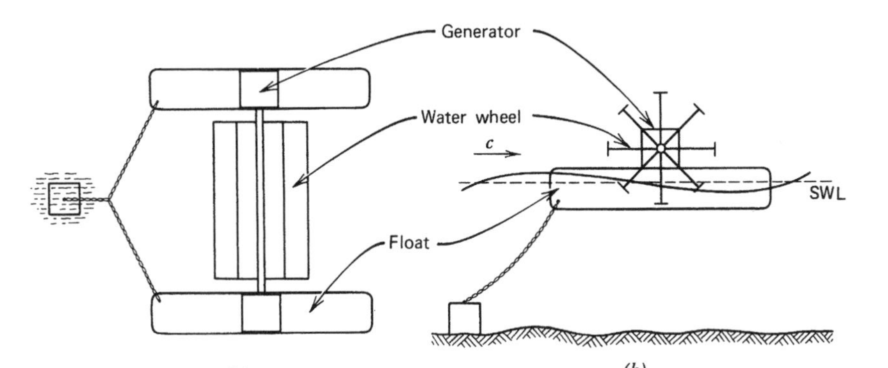

Figure 4.21 Sketch of a floating water wheel wave energy convertor: (*a*) top view; (*b*) side view (cutaway).

to be much less than the wavelength λ. Thus the dynamic pressure acting on the blades will vary slightly over D_p, thus resulting in little energy transfer to the wheel. For the case of D_p with the same order of magnitude as λ, either the waves are of low energy for *short waves* or the capital costs of the water wheel are prohibitive for *long waves*.

For the optimal design where the axis of rotation is at $z = +H/2$, ideally the water wheel will make one-half a revolution per wave; that is, the average *rotational velocity* will be

$$\omega_p = \frac{\omega}{2} = \frac{\pi}{T} \tag{4.69}$$

in radians per second. The product of the *resisting torque* of the generator T_p and the rotational velocity ω_p results in the power converted by the device; thus

$$\begin{aligned} P_p &= T_p \omega_p \\ &= \frac{T_p \pi}{T} \end{aligned} \tag{4.70}$$

Further mathematical analysis of the water wheel is not justified here since it is evident that the rotational velocity of the wheel decreases as the period increases, and the wave power is greater for the higher-period waves.

Another popular particle motion device is the bottom-mounted *compliant flap* sketched in Figure 4.22. The compliant flap is designed to respond to the dynamic pressures of the particle motions. Thus, below a depth of $\lambda/2$ where practically no motions exist, the flap remains in a vertical orientation. This vertical orientation is maintained either by buoyancy or some internal spring. The *excursion* of the top of the flap l_f will be less than the maximum particle excursion in the horizontal direction. In *deep water* the particles on the surface of the water travel a distance equal to the wave height H over one period, according to the linear theory described in Section 2.1; thus $l_f < H$.

The horizontal particle velocity in deep water is described by equation (2.12):

$$u = \frac{\pi H}{T} e^{kz} \cos(kx - \omega t) \tag{2.12}$$

where the wave number is

$$k = \frac{2\pi}{\lambda} \tag{2.41}$$

To minimize "feedback" effects of the flap motions to the waves, we design the *breadth* of the flap B_f to be much less than the wavelength. For short waves (i.e., $\lambda \simeq B_f$), the waves will mostly be reflected and little wave energy

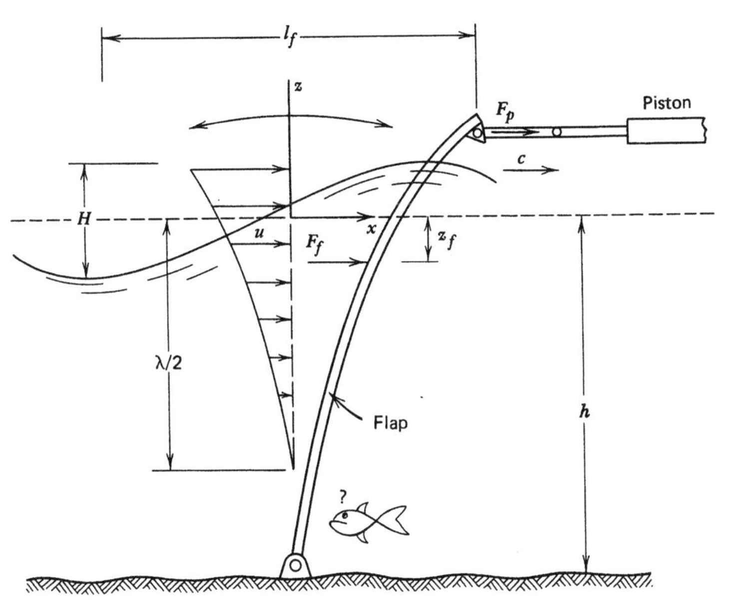

Figure 4.22 Schematic diagram of a compliant flap-wave energy conversion system in deep water.

absorbed by the flaps. The wave-induced *dynamic pressure* on the flap is obtained from equation (4.48):

$$p = -\rho \frac{\partial \varphi}{\partial t}$$

$$= \frac{\rho g H}{2} e^{kz} \cos(\omega t) \tag{4.48}$$

where the neutral or calm water position of the flap is at $x = 0$. The *force* due to this pressure is then

$$F_f = B_f \int_{-h}^{0} p_d \, dz$$

$$= \frac{B_f \rho g H}{2} \cos(\omega t) \int_{-h}^{0} e^{kz} \, dz$$

$$= \frac{B_f \rho g H}{2k} \cos(\omega t) \left[1 - e^{-kh} \right]$$

$$\simeq \frac{B_f \rho g^2 T^2 H}{8\pi^2} \cos(\omega t) \tag{4.71}$$

The last approximation is valid since $h > \lambda/2$ in deep water. The *center of pressure*, that is, the point on the flap at which F_f acts, is at a depth obtained from

$$\bar{z}_f = \frac{\int_{-h}^{0} p_d z \, dz}{\int_{-h}^{0} p_d \, dz} = \frac{-1}{k} = \frac{-\lambda}{2\pi} \tag{4.72}$$

as shown in Figure 4.22.

The force transmitted to the *hydraulic piston* sketched in Figure 4.22 is found by taking moments about the depth $z = -\lambda/2$. Thus

$$F_p \frac{\lambda}{2} = F_f\left(\frac{\lambda}{2} + \bar{z}_f\right)$$

or, using the results of equations (4.71) and (4.72),

$$F_p = F_f\left(1 - \frac{1}{\pi}\right) = \frac{B_f \rho g^2 T^2 H}{8\pi^2}\left(1 - \frac{1}{\pi}\right)\cos(\omega t) \tag{4.73}$$

The *total energy* transferred to the piston over one wave period is

$$E_f = \frac{4}{T}\int_0^{T/4} F_p l_f \cos(\omega t)\, dt = \frac{B_f \rho g^2 T^2 H}{4\pi^3}\left(1 - \frac{1}{\pi}\right) l_f \tag{4.74}$$

where, again, $l_f < H$.

Example 4.10

Consider a bottom-mounted compliant flap located in 50 ft (15.2 m) of water and subject to a deep water wave of height 3 ft (0.914 m) and period 4 sec. Using the results of equation (2.6), the wavelength of a 4-sec deep water wave is derived as 82.0 ft (25.0 m). Little wave action, therefore, is experienced below a depth of $\lambda/2 = 41.0$ ft (12.5 m). Referring to Figure 4.22, only the upper 41.0 ft (12.5 m) of the flap will respond to the wave. The stroke of the piston l_f is 1.5 ft (0.457 m), and the breadth of the flap B_f is 10 ft (3.05 m); thus the energy converted by the system is obtained from equation (4.74):

$$E_f = \frac{B_f \rho g^2 T^2 H}{4\pi^3}\left(1 - \frac{1}{\pi}\right) l_f$$

$$= \frac{10(2.00)(32.2)^2 4^2(3)}{4\pi^3}\left(1 - \frac{1}{\pi}\right)1.5$$

$$= 8200 \text{ ft-lb } (11{,}100 \text{ J})$$

From the results of equation (2.17), the total energy of the wave is

$$E = \frac{\rho g H^2 \lambda B_f}{8}$$

$$= \frac{2.00(32.2)3^2 82.0(10)}{8}$$

$$= 59{,}400 \text{ ft-lb } (80{,}600 \text{ J})$$

the *efficiency* of the compliant flap in this case is then

$$\epsilon_f = \frac{E_f}{E}$$

$$= \frac{4(1 - 1/\pi)}{\pi^2}\frac{l_f}{H}$$

$$= 0.138 \quad \text{or} \quad 13.8\% \qquad (4.75)$$

From the discussion of the water wheel wave energy convertor, it can be assumed that this device is not feasible because of its apparent inefficiency. Furthermore, it is sensitive to wave direction; thus waves traveling in a direction parallel to the rotational axis will not excite the device. The water wheel system is cumbersome and, therefore, is costly to construct, moor, and maintain.

The results of equation (4.75) show that the maximum theoretical efficiency of the compliant flap corresponds to a stroke l_f equal to the wave height H, and the maximum theoretical efficiency is 27.6%. The reader may wonder why the maximum theoretical efficiency is not 50%, that is, equal to the kinetic energy of the wave; the reason is that it is practically impossible to have the flap exactly follow the particle motions and still deliver usable energy to the hydraulic system. The flap in shallow water is more feasible since the horizontal motions of the water particles do not vary significantly from the free surface to the seafloor. Furthermore, the capital costs in shallow water will be much less than those in deep water. Unfortunately, the total wave energy in shallow water is significantly less than that in deep water due to both friction and the irrotationalities of shoaling.

4.2 Advanced Techniques

The basic techniques described in Section 4.1 are those that have been suggested and patented in one form or another since the turn of the twentieth century. More advanced techniques of wave energy conversion that have been proposed over the past few decades are described in this section. The first three of these techniques, the *nodding duck*, the *contouring raft*, and the *Russell rectifier*, originated in the United Kingdom. These are described in the book by Ross (1979), which is a historical account of wave energy conversion in the United Kingdom. The other advanced techniques are grouped together under the title "Wave Focusing." These wave focusing techniques allow a relatively small wave energy conversion device to capture energy from a wave crest width that may be many times the width of the device. The focusing techniques include the *antenna effect* of resonating bodies, *lens focusing* by submerged structures, and focusing by *refraction* of semisubmerged structures. These focusing techniques are of major interest in the United States.

In the sections that follow a basic physical description of each advanced technique is presented. This description is then followed by a simplified mathematical analysis of the performance. Finally, examples of the design and the performance of each device are presented.

A Salter's Nodding Duck

In his 1974 paper Salter introduced a unique wave energy conversion concept capable of efficiencies approaching 90% in two-dimensional sinusoidal waves. He referred to his device as the "nodding duck" because of both its profile and operation, illustrated in Figure 4.23. Referring to Figure 4.23*a*, the "*paunch*" of the duck is shaped such that the dynamic pressure caused by the wave-induced water particle motions efficiently force the duck to rotate about an axis through point O. In addition to the dynamic pressure, the changing hydrostatic pressure contributes to the rotation by causing the buoyant forebody near the "beak" to rise and fall. Since both of these pressure-induced motions are in phase, the nodding duck converts both the *kinetic* and *potential* energies of the wave into *rotational mechanical energy*. The rotational motion is then converted into electrical energy by a hydraulic–electric subsystem. A mathematical analysis of the operation is found in the paper by Salter et al. (1976). This analysis has been extended by Mynett et al. (1979).

As for most wave energy conversion devices, we begin our analysis of the

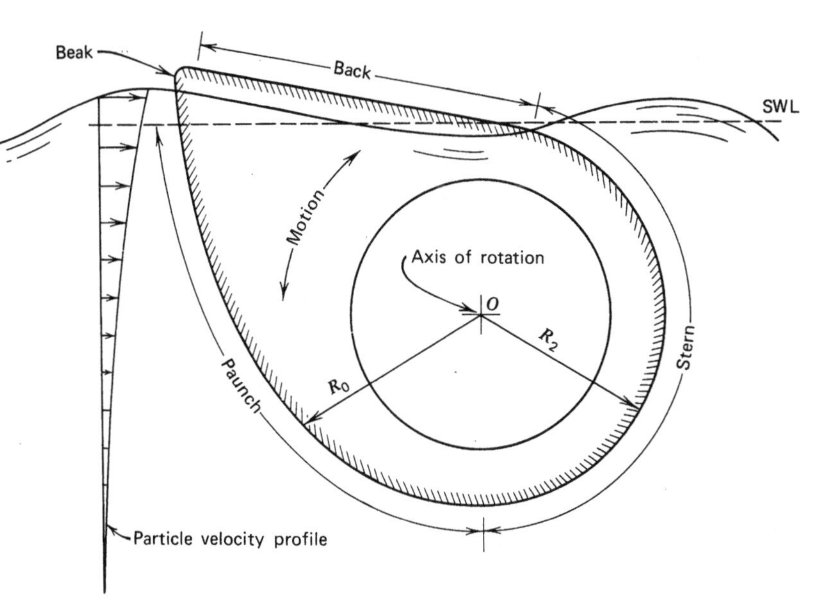

Figure 4.23*a* Nomenclature for the nodding-duck-wave energy convertor.

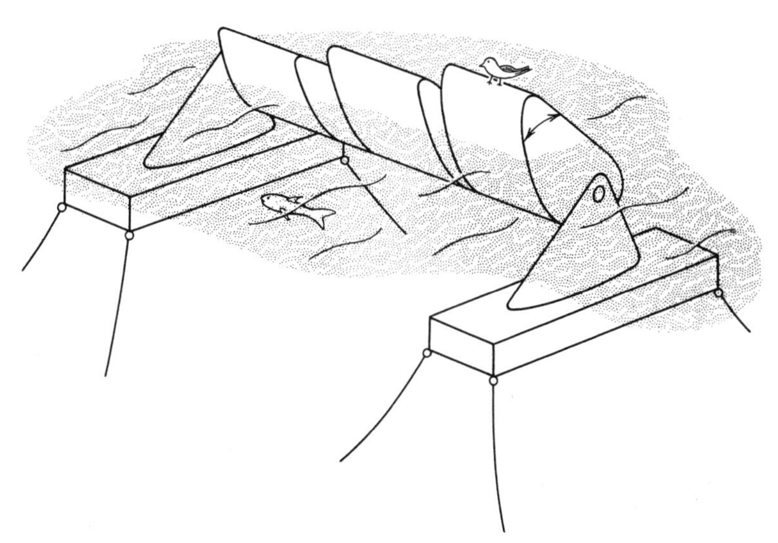

Figure 4.23*b* Sketch of a moored three-duck system.

nodding duck by specifying a *design wave* that corresponds to the *natural frequency* of the system. In the most general terms, this natural frequency is defined by

$$f_D = \frac{1}{2\pi} \sqrt{\frac{\text{stiffness}}{\text{inertia}}} \tag{4.76}$$

where the stiffness and the inertia of the system are to be determined. Thus the *design condition* is obtained by combining the frequency expressions of equations (2.2) and (4.76); thus

$$\begin{aligned} f_D &= f \\ &= \frac{1}{2\pi}\sqrt{\frac{2\pi g}{\lambda}\tanh\left(\frac{2\pi h}{\lambda}\right)} \end{aligned} \tag{4.77}$$

To simplify the analysis, assume that the duck operates in *deep water*. Thus equation (4.77) is approximated by

$$f_D \simeq \sqrt{\frac{g}{2\pi\lambda}} \tag{4.78}$$

where λ is the wavelength.

The average deep water wavelength is known from *in situ* measurements made over one or more years. With this information, Salter et al. (1976) then specify that the paunch "radius," R_0 in Figure 4.23*a*, is

$$R_0 = K_D e^{2\pi z/\lambda} \tag{4.79}$$

where K_D is a design constant that is determined at $z = -2R_2$ for an assumed value of λ/R_2. For the results obtained by Salter et al., (1976) $\lambda/R_2 = 20$ is assumed. With this condition, equation (4.79) becomes

$$\begin{aligned} R_0\big|_{z=-2R_2} &= K_D e^{-\pi/5} \\ &= 0.533 K_D \\ &= \frac{\lambda}{20} \end{aligned}$$

Thus the design constant is

$$K_D = 0.0937\lambda \tag{4.80}$$

Example 4.11

The design wave at a deep water location in the North Pacific is found to have a period T of 10 sec and a height H of 5 ft (1.52 m). From the results of equation (2.6), the corresponding wavelength is

$$\begin{aligned} \lambda &= \frac{gT^2}{2\pi} \\ &= \frac{32.2(10)^2}{2\pi} \\ &= 512 \text{ ft } (156 \text{ m}) \end{aligned}$$

Assuming $\lambda / R_2 = 20$, as in the derivation of equation (4.80), the "stern" radius R_2 is 25.6 ft (7.80 m). The *draft* of the duck is then

$$d = 2R_2$$
$$= 51.2 \text{ ft } (15.6 \text{ m})$$

The equation for the paunch "radius" is obtained from the combination of equations (4.79) and (4.80):

$$R_0 = 0.0937\lambda e^{2\pi z/\lambda}$$
$$= 48.0e^{0.0123z} \text{ (ft)}$$

or

$$R_0 = 14.6e^{0.0403z} \text{ (m)}$$

In deep water no significant water particle motions occur below $z = -\lambda/2 = -256$ ft (78.0 m), and the duck in this situation is then exposed to the upper 20% of the water particle motions. From the results found in McCormick (1973) and elsewhere, the *power* in the linear deep water wave from $z = -d$ to the free surface is obtained from the expression

$$P_D = \frac{\rho g H^2 c B}{16}(1 - e^{-2kd}) \tag{4.81}$$

The *total power* within the deep water wave is obtained from equation (2.19), where the group velocity c_g is half of the phase velocity c; thus

$$P = \frac{\rho g H^2 c B}{16}$$

The *percentage* of wave power available to the nodding duck is, then,

$$\frac{P_D}{P} = (1 - e^{-2kd})$$
$$= 1 - e^{-4\pi(51.2)/512}$$
$$= 0.715 \quad \text{or} \quad 71.5\% \tag{4.82}$$

Returning to the frequency expression given in equation (4.76), Salter et al. (1976) show that the *effective stiffness* of the nodding duck system is

$$\text{Stiffness} = \tfrac{1}{3}\rho g L_D^3 B - M g r_D \cos(\Lambda) \tag{4.83}$$

where M is the rotating mass, r_D is the location of the center of mass with respect to the axis of rotation, and Λ is the resting angle of r_D with respect to the horizontal plane. The *inertial term* of equation (4.76) is (approximately)

$$\text{Inertia} = I_y + \frac{1}{32}\pi\rho B\left[\left(\frac{L_D}{\cos(\Lambda)}\right)^2 - R_2\right]^2 \tag{4.84}$$

where I_y is the mass moment of inertia of the system about the rotational axis. It should be noted here that I_y depends on the distribution of the mass of the duck body including any internal energy conversion mechanisms.

Combining the expressions of equations (4.83) and (4.84) with that of equation (4.76), the *natural frequency* of the system is

$$f_D = \frac{1}{2\pi}\sqrt{\frac{\frac{1}{3}\rho g L_D^3 B - Mgr_D \sin(\Lambda)}{I_y + \frac{1}{32}\pi\rho B\left[\left(\frac{L_D}{\cos(\Lambda)}\right)^2 - R_2\right]^2}}$$
$$= \frac{1}{T_D} \tag{4.85}$$

where T_D is the natural period.

***Note*:** In the added-mass moment of inertia expression given in equations (4.84) and (4.85), we have assumed that the extension of the vector r_D intersects the SWL at the paunch profile, referring to Figure 4.24.

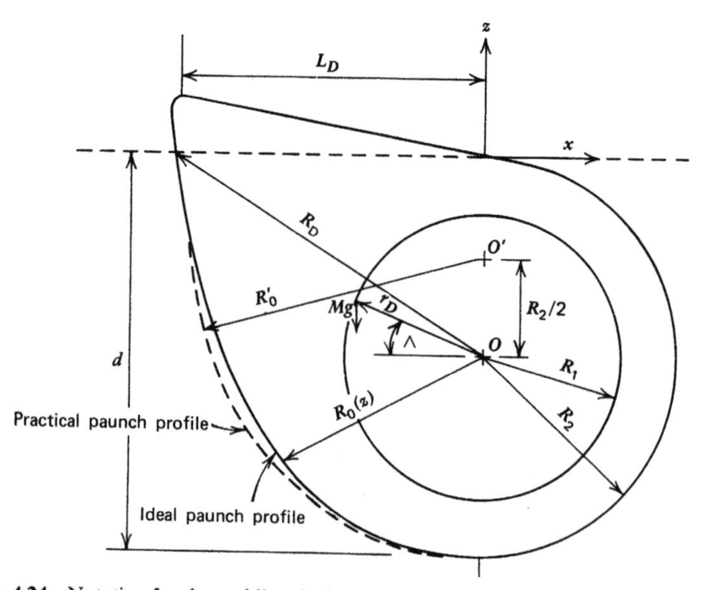

Figure 4.24 Notation for the nodding-duck profile with ideal and practical paunch profiles.

Example 4.12

For the nodding duck described in Example 4.11, assume that the center of mass is located such that r_D and R_D coincide, where $r_D = 0.3R_D$. Referring to Figure 4.24, we see that the water plane length is

$$\begin{aligned} L_D &= R_D \cos(\Lambda) \\ &= \sqrt{R_D^2 - R_2^2} \end{aligned} \tag{4.86}$$

where $R_2 = 25.6$ ft (7.80 m) and $R_D = R_0|_{z=0} = 48.0$ ft (14.6 m). Equation (4.86) now yields

$$\begin{aligned} L_D &= (48.0)^2 - (25.6)^2 \\ &= 40.6 \text{ ft } (12.4 \text{ m}) \end{aligned}$$

and

$$\begin{aligned} \Lambda &= \cos^{-1}\left(\frac{L_D}{R_D}\right) \\ &= 32.2° \end{aligned} \tag{4.87}$$

We assume in this example and also in Example 4.13 that the mass of the system M is equal to the mass of the displaced water, although this is not necessarily so, according to Salter et al. (1976). Before calculating the value of the mass, we *modify the ideal design* of Salter et al. (1976). The reasons for this modification are to (a) simplify the calculation of the displaced mass and (b) facilitate the fabrication of the nodding duck. The modification is accomplished by replacing the paunch curve, described by R_0 in equation (4.79), by a circular arc of radius

$$\begin{aligned} R_0^1 &= 1.5R_2 \\ &= 38.4 \text{ ft } (11.7 \text{ m}) \end{aligned} \tag{4.88}$$

with its origin O^1 at $z = -R_2/2 = -12.5$ ft (3.90 m) as shown in Figure 4.24. The curve of the modified paunch extends over 80°. Between the upper end of the arc and the beak is a line tangential to the arc. Using this modified geometry, and assuming the width of the duck to be $B = 2R_2 = 51.6$ ft (15.7 m), the displaced mass is

$$\begin{aligned} M &= \rho B \left\{ \frac{\pi R_2^2}{2} + 0.222\pi\left(R_0^1\right)^2 + 0.5\left(R_0^1\right)^2 \sin(10^0)\left[\cos(10^0) + \tan(10^0)\right] \right. \\ &\quad \left. + \frac{R_2 L_D}{2} - \frac{R^2}{8^2}\tan(10^0) \right\} \\ &= 2.00(51.2)\{1030 + 1030 + 148 + 520 - 14.4\} \\ &= 2.78 \times 10^5 \text{ lb-ft/sec } (4.05 \times 10^6 \text{ kg}) \end{aligned} \tag{4.89}$$

We now have enough information to determine the natural frequency of the system if we assume that the mass moment of inertia I_y is equal to the added-mass moment of inertia, that is, the second term on the right hand side of equation (4.84). Equation (4.85) now yields

$$
\begin{aligned}
f_D &= \frac{1}{2\pi}\sqrt{\frac{0.333\rho g L_D^3 B - Mg(0.3R_D)\sin(\Lambda)}{2\left\{\frac{1}{32}\pi\rho B\left[\left(\frac{L_D}{\cos(\Lambda)}\right)^2 - R_2^2\right]^2\right\}}} \\
&= \frac{1}{2\pi}\sqrt{\frac{74.1\times 10^6 - 68.7\times 10^6}{54.7\times 10^6}} \\
&= 0.0500 \text{ Hz}
\end{aligned} \tag{4.90}
$$

or

$$T_D = 20.0 \text{ sec}$$

which, from equation (2.6), corresponds to a wavelength of 2050 ft (625 m). Thus the natural period T_D and the wave period T are not equal; the latter is 10 sec. To make these two periods equal, we simply change the position of the center of mass from $r_D = 0.3R_D$ to $r_D = 0.229R_D = 11.0$ ft (3.35 m).

From the results in Example 4.12 we see that in the design of the nodding duck wave energy convertor provision should be made to adjust the position of the center of mass to match the natural period T_D and the wave period T. This adjustment involves changing the position vector r_D and not the angle Λ.

The performance of the nodding duck operating in two-dimensional sinusoidal waves is analyzed by Mynett et al. (1979). This analysis includes the effects of mooring flexibility on the efficiency. The results of the study show that an efficiency of 100% can be attained under ideal operating conditions. Salter et al. (1976) actually attained an energy conversion efficiency of 89% for a small-scale model under similar wave conditions. The analysis of Mynett et al. (1979) has been extended by Serman (1978) to predict the performance of the duck in a random sea.

Significant experimental data were obtained by Carmichael (1978) in wave tank tests of a small-scale model. Operating under two-dimensional sinusoidal wave conditions, Carmichael confirmed the high-efficiency performance observed by Salter et al. (1976). Then Carmichael conducted tests on the same model in *three-dimensional waves* and attained *conversion efficiencies up to 150%* for wave fronts aligned with the axis of rotation of the duck. The efficiency calculation was based on the available *two-dimensional wave power*;

that is, from equation (2.19) where b is replaced by B,

$$P = \frac{\rho g H^2 c_g B}{8} \tag{4.91}$$

The reader must now wonder how an efficiency greater than 100% can be attained. There are two possible answers. First, it is possible that reflections from the wave tank walls occurred; thus from the results presented in Section 3.2, a significant increase in wave height would be observed. From the data presented by Carmichael (1978), this possibility is discounted since the phase relation between the incident and reflected wave systems would not always be optimal in the region of the model. The second possible cause of the abnormally high conversion efficiencies is *wave focusing* by the "antenna effect" described in Section 4.2, D. This focusing results from the interaction between the incident waves and the waves radiated from the body at or near a condition of *resonance*, that is, when $T_D \simeq T$. Wave power is focused on the body from a crest width b, which may be many times greater than the width B of the nodding duck. The author believes that wave focusing is the actual cause of the high efficiencies observed by Carmichael (1978), whose data are presented in dimensionless form in Figure 4.25, where the efficiency ϵ is presented as a function of the *dimensionless frequency* $2\pi f\sqrt{R_2/g}$.

A second significant finding of Carmichael (1978) concerns the effect of *mooring flexibility*. Referring to the curves in Figure 4.26, it can be seen that the efficiency ϵ is lowest under the slack mooring condition and highest (over most frequencies) when the duck model is rigidly positioned. This is expected

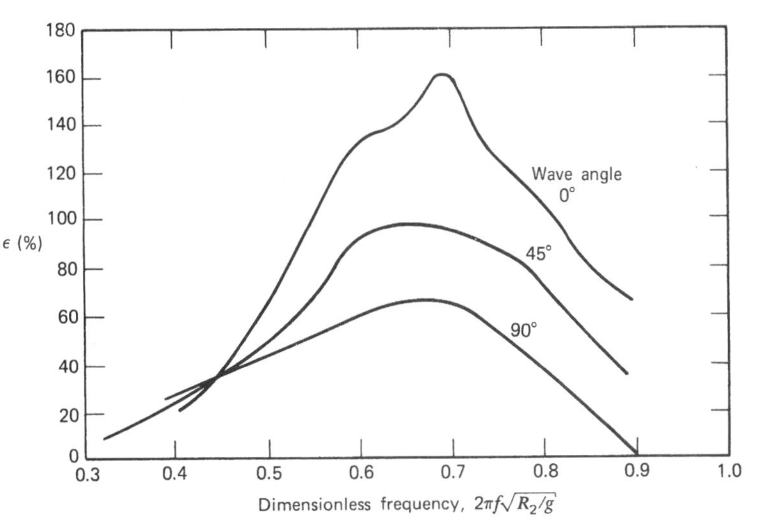

Figure 4.25 Duck efficiency as a function of frequency for three wave angles. From Carmichael (1978).

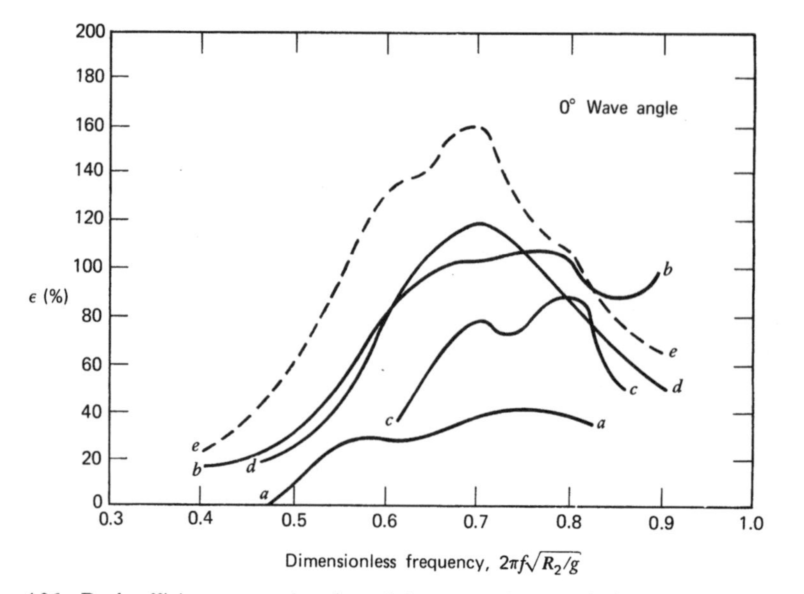

Figure 4.26 Duck efficiency as a function of frequency for (*a*) slack moorings, (*b*) chain taut-moorings, (*c*) single taut-moorings, (*d*) four-chain taut moorings, and (*e*) rigid frame. From Carmichael (1978).

since the slack mooring allows part of the wave energy to be absorbed by degrees of freedom of duck motion that do not contribute to the energy conversion. The rigid mount allows only the design energy conversion motions to occur.

Example 4.13

Consider the nodding duck described in Examples 4.11 and 4.12 to be taut moored in deep water sinusoidal waves that directly approach the duck. This situation corresponds to curve d in Figure 4.26. The duck is designed to resonate in a 10-sec wave; thus

$$T_D = T = 10 \text{ sec}$$

that has a wave height H of 5 ft (1.52 m). The *optimum* (*design*) *condition* for the four-chain taut mooring in Figure 4.26 occurs when

$$2\pi f\sqrt{\frac{R_2}{g}} = 0.7$$

For the prototype described in Examples 4.11 and 4.12, the dimensionless frequency is

$$\frac{2\pi}{T_D}\sqrt{\frac{R_2}{g}} = \frac{2\pi}{10}\sqrt{\frac{25.6}{32.2}}$$

$$= 0.560$$

Thus, from curve *d* in Figure 4.26, the efficiency is approximately 60%, which is about half of the maximum of curve *d*. To obtain this maximum efficiency, the design wave period should be 8.00 sec. This period can be obtained, as in Example 4.12, by adjusting the position of the center of mass, that is, by adjusting r_D. The new design value condition is

$$r_D = 0.176 R_D$$

$$= 8.46 \text{ ft } (2.58 \text{ m})$$

for the design geometry. These values do show how the *design period is sensitive to the position of the center of mass.*

Using the preceding value of r_D corresponding to the 10-sec period, the efficiency value is 60%. Thus for the 5-ft (1.524-m), 10-sec wave, the power converted by the 51.6-ft (15.7-m)-wide nodding duck is

$$P_D = 0.60 P$$

$$= 0.6 \frac{\rho g H^2 (\lambda / T) B}{16}$$

$$= 0.6 \frac{(2.00) 32.2 (5)^2 (512/10) 51.6}{16}$$

$$= 1.61 \times 10^5 \text{ ft-lb/sec } (218 \text{ kW})$$

The results in Example 4.13 indicate that the *optimum design* of the nodding duck, using the data due to Carmichael (1978), is more difficult to obtain than previously expected. The primary reason for this difficulty is the sensitivity of the design period T_D as a function of r_D. In the final design of the duck a *feedback system* is required to readjust r_D such that the design condition (i.e., $T_D = T$) is always met.

Carmichael (1978) also performed tests in *random waves*, and his results are shown in Figure 4.27. In that figure the effects of mooring rigidity are also shown. Again, the more rigid mooring results in a greater wave energy conversion efficiency. Comparing the results in Figure 4.27 with curves *d* and

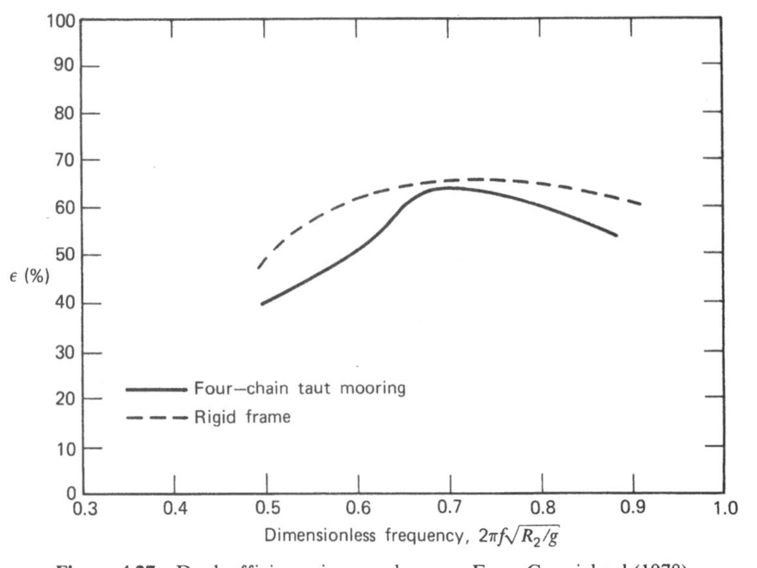

Figure 4.27 Duck efficiency in a random sea. From Carmichael (1978).

Plate 4.3 An artist's concept of a salter nodding-duck-wave energy conversion system. Courtesy of the Energy Technology Support Unit, AERE Harwell—The United Kingdom.

e in Figure 4.26, it can be seen that the system is far less efficient in random seas.

The *costs of anchoring and mooring* were found to be dominant in the United Kingdom Wave Energy Conversion Symposium held at Heathrow Airport in November 1978. These costs for the nodding duck prototype were projected to be such that the delivered power is far from competitive in the 1978–1979 energy market. This cost projection caused the United Kingdom to reassess its wave energy conversion program. Carmichael (1978) also predicted high delivered power costs based on the capital costs of the nodding duck. For a five-duck system operating off the Northeast Coast of the United States, Carmichael predicts a delivered power cost of $4305/kW (in 1978 dollars) for a total power delivery of 5715 kW. Thus although the power conversion efficiency of the nodding duck is high, the capital costs of the system must be significantly reduced before this method of wave energy conversion becomes competitive.

B Cockerell's Rafts

Sir Christopher Cockerell, inventor and developer of the Hovercraft, suggested using wave *contouring rafts* as a means of converting the energy of waves. These rafts are hinged together, and an energy conversion subsystem (usually hydraulic) is located at each hinge. A historical account of Cockerell's efforts is found in the book by David Ross (1979). Cockerell's first scheme involved rafts of equal lengths. A variation of this idea was suggested and tested by Glenn Hagen in the United States where rafts of progressively increasing lengths were used. Since the ideas of Hagen and Cockerell were independent and almost simultaneous, the contouring rafts are sometimes referred to as the *Hagen–Cockerell rafts*. Within a short period of time after his original idea Cockerell redesigned his contouring rafts to include varying raft lengths.

A comprehensive theoretical study of the contouring rafts has been performed by Pierre Haren (1978) for his master's thesis at the Massachusetts Institute of Technology (MIT) under the guidance of Professor C. C. Mei. Haren's study consists of analyses of (*a*) a single raft hinged at a wall in long waves, (*b*) multiple rafts of varying lengths in shallow water, and (*c*) three-raft systems in both shallow and deep water waves. In all cases the sea is assumed to be *two dimensional*. The discussion of the contouring rafts presented herein is based on Haren's work. Attention is focused on the results of the theoretical study rather than on the methodology since the analysis, although not overly complicated, is far too elaborate for presentation herein. Particular attention is paid to the *three-raft system* since this appears to be the most cost effective.

A sketch of a three-raft system is shown in Figure 4.28*a*, and schematic diagrams of the system are presented in Figure 4.28*b* (calm water) and Figure 4.28*c* (sinusoidal waves). The energy conversion of the contouring rafts depends on the *relative angular motions of raft pairs.* These angular motions are

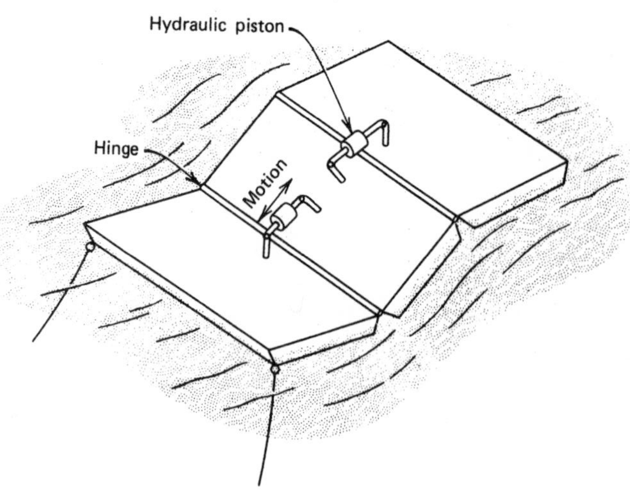

Figure 4.28*a* Sketch of a contouring raft system.

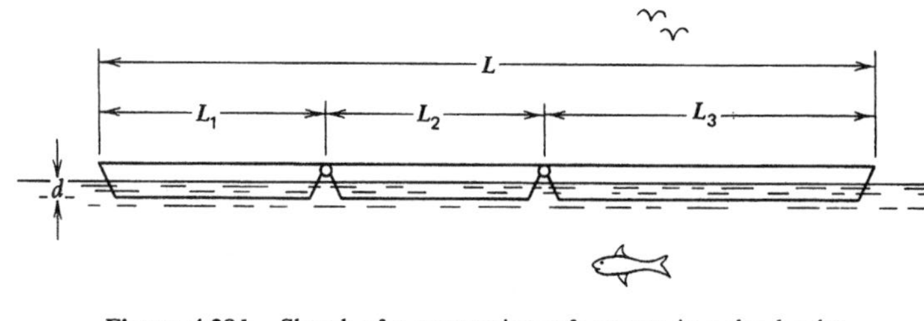

Figure 4.28*b* Sketch of a contouring raft system in a dead calm.

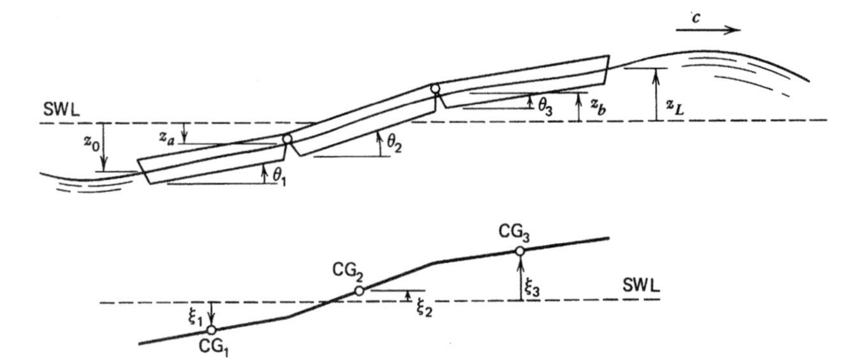

Figure 4.28*c* Sketch of a contouring raft system in waves (upper), with center of gravity (CG) displacements (lower).

described by the following coupled equations of motion:

$$(I_1 + I_{w1})\ddot{\theta}_1 + \alpha_a(\dot{\theta}_1 - \dot{\theta}_2) + \frac{R_a L_1}{2} = M_1 \circlearrowleft^{+} \quad (4.92)$$

$$(I_2 + I_{w2})\ddot{\theta}_2 + \alpha_a(\dot{\theta}_2 - \dot{\theta}_1) + \alpha_b(\dot{\theta}_2 - \dot{\theta}_3) + \frac{(R_a + R_b)L_2}{2} = M_2 \circlearrowleft^{+} \quad (4.93)$$

$$(I_3 + I_{w3})\ddot{\theta}_3 + \alpha_b(\dot{\theta}_3 - \dot{\theta}_2) + \frac{R_b L_3}{2} = M_3 \circlearrowleft^{+} \quad (4.94)$$

where the "dots" over the θ values indicate time differentiation and where I_j $(j = 1, 2, 3)$ is the mass moment of inertia of raft j with respect to its center, I_{wj} is the added-mass moment of inertia, α_a and α_b are the *energy extraction rates* at each hinge, $R_a(t)$ and $R_b(t)$ are the vertical reaction forces at each hinge, and $M_j(t)\circlearrowleft^{+}$ is the wave-induced moment about the center of raft j.

***Note*:** The center of gravity of each raft is located in the water plane at the geometric center. The reaction forces $R_a(t)$ and $R_b(t)$ are unknowns.

A similar set of equations can be written for the linear vertical motions (*heave*) of the raft system. Referring to raft 3 in Figure 4.28*c* for illustration, the linear displacement of the center of that raft is

$$\xi_3 = \frac{z_b + z_L}{2} \quad (4.95)$$

and similarly for rafts 1 and 2. The *coupling* of the *rotational* and *rectilinear* motions for raft 3 is through the relationship

$$\sin(\theta_3) \simeq \theta_3$$

$$= \frac{z_L - z_b}{L_3}$$

$$= \frac{A_L - A_b}{L_3} e^{j\omega t} \quad (4.96)$$

where all angular displacements are assumed to be small. The *wave frequency*

$$\omega = 2\pi f$$

$$= \frac{2\pi}{T} \quad (4.97)$$

must be that of the raft system for that system to be *contouring*. The amplitudes A_L and A_b are complex numbers since we have chosen the complex time function

$$e^{j\omega t} = \cos(\omega t) + j\sin(\omega t) \tag{4.98}$$

to represent both wave and raft motions.

We now have a *system of six equations* and *six variables* θ_i and ξ_i $(i = 1, 2, 3)$, that with certain specific assumptions can be solved simultaneously. Our intention, however, is to present results of the analysis and not the analysis itself. Those readers interested in the analytical technique should read the thesis of Haren (1978), who solves the equations of motion numerically.

The *energy* converted by the contouring raft system undergoing linear displacements [i.e., $\sin(\theta_j) \simeq \theta_i$, etc.] is obtained from

$$\begin{aligned}
\hat{E}_H &= \frac{B}{T}\int_0^T \left[\alpha_a(\dot{\theta}_2 - \dot{\theta}_1)^2 + \alpha_b(\dot{\theta}_3 - \dot{\theta}_2)^2\right] dt \\
&= \frac{4B}{T}\int_0^{T/4} \mathrm{Re}\left\{ \alpha_a \left[\frac{(A_b - A_a)}{L_2} - \frac{(A_a - A_0)}{L_1}\right]^2 e^{j\omega t} \right. \\
&\quad \left. + \alpha_b \left[\frac{(A_L - A_b)}{L_3} - \frac{(A_b - A_a)}{L_2}\right]^2 e^{i\omega t} \right\} dt \\
&= \frac{2B}{\pi} \mathrm{Re}\left\{ j\alpha_a \left[\frac{(A_b - A_a)}{L_2} - \frac{(A_a - A_0)}{L_1}\right]^2 \right. \\
&\quad \left. + j\alpha_b \left[\frac{(A_L - A_b)}{L_3} - \frac{(A_b - A_a)}{L_2}\right]^2 \right\}
\end{aligned} \tag{4.99}$$

The *efficiency* of the energy conversion is then

$$\begin{aligned}
\epsilon &= \frac{\hat{E}_H}{E} \\
&= \frac{16}{\pi\rho g H^2 \lambda} \mathrm{Re}\left\{ j\alpha_a \left[\frac{(A_b - A_a)}{L_2} - \frac{(A_a - A_0)}{L_1}\right]^2 \right. \\
&\quad \left. + j\alpha_b \left[\frac{(A_L - A_b)}{L_3} - \frac{(A_b - A_a)}{L_2}\right]^2 \right\}
\end{aligned} \tag{4.100}$$

based on a two-dimensional wave having a crest length equal to the raft width. To utilize equations (4.99) and (4.100), we must first determine the amplitudes of motion A_0, A_a, A_b, and A_L by solving the six equations of motion. In addition, the reaction forces at the hinges R_a and R_b must also be determined. As previously mentioned, Haren (1978) performs these tasks using *numerical techniques* and, in addition, performs an *optimization study* of the contouring raft system. The results of the optimization study are now presented. Of interest are the following four *design conditions* that Haren considered:

1 Optimized design based on the wave at the peak of the energy spectrum assuming a value of the total length

$$L = L_1 + L_2 + L_3 \tag{4.101}$$

2 Optimized design based on the wave at the peak of the energy spectrum and determining the most cost-effective length.

3 Optimized design based on a double peaked spectrum assumming a total length L.

4 Optimized design based on a double peaked spectrum and determining the most cost-effective length.

Results of Haren's (1978) optimization studies are contained in Examples 4.14 through 4.17.

Example 4.14

Results of condition 1 are obtained assuming a total raft length of 450 ft (137 m). The raft system is assumed to be located in 30 ft (9.14 m) of water where the design wave height and period are $H = 2$ ft (0.610 m) and $T = 5.52$ sec, respectively, corresponding to the *peak of the energy spectrum*. The wavelength in this situation is $\lambda = 171$ ft (52.1 m). With these assumptions the *optimized raft lengths* are

$$L_1 = 54.8 \text{ ft } (16.7 \text{ m})$$

$$L_2 = 171 \text{ ft } (52.1 \text{ m})$$

$$L_3 = 224 \text{ ft } (68.3 \text{ m})$$

with *maximum angular differences* of

$$|\theta_1 - \theta_2| = 0.033 \text{ rad}(1.89°)$$

and

$$|\theta_2 - \theta_3| = 0.003 \text{ rad}(0.172°)$$

The efficiency ϵ_D for this optimized system is presented in Figure 4.29 as

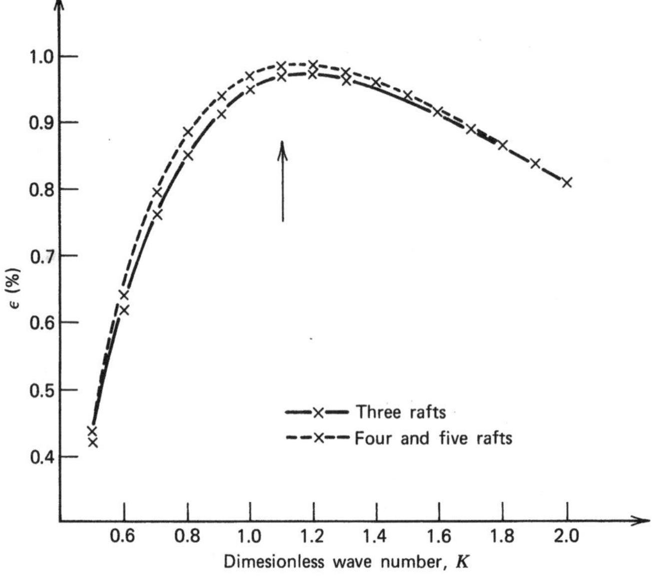

Figure 4.29 Raft efficiency as a function of wave number for peak frequency optimization (↑). From Haren (1978). The raft system length is assumed.

a function of the *dimensionless wave number*

$$K = kh = \frac{2\pi h}{\lambda} \tag{4.102}$$

The *peak efficiency* is 96%. Although our primary interest is in a three-raft system, Haren's (1978) results in Figure 4.29 show that a slight increase in efficiency can be obtained by using four and five rafts over the given system length L. The increased number of rafts, however, significantly adds to the *capital costs*. Thus the increased efficiency does not warrant the increased costs of the four- and five-raft systems.

Example 4.15

For condition 2, using the peak spectrum wave having a 2-ft (0.610-m) height and a 5.52-sec period, the optimum design lengths are

$$L_1 = 48.0 \text{ ft } (14.6 \text{ m})$$

$$L_2 = 116 \text{ ft } (35.4 \text{ m})$$

$$L_3 = 50.4 \text{ ft } (15.4 \text{ m})$$

Thus the total length of the three-raft system is 214 ft (65.3 m), which is

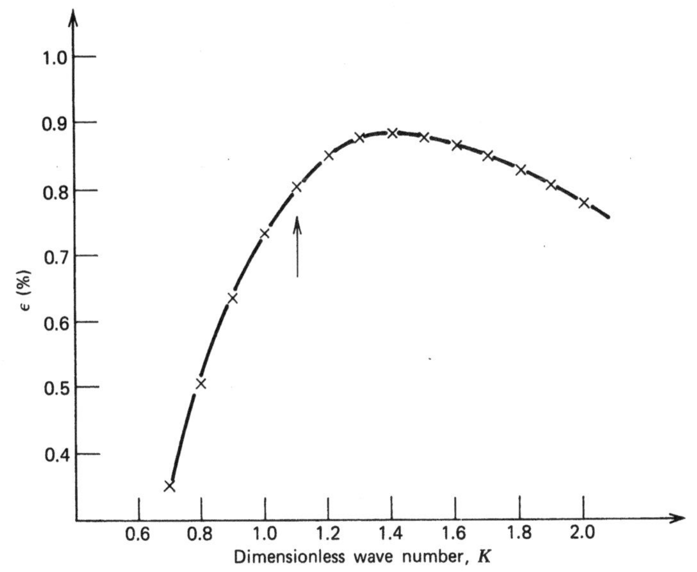

Figure 4.30 Raft efficiency as a function of wave number for peak frequency optimization (↑) for three- and four-raft systems. From Haren (1978).

less than half of the system obtained for condition 1 immediately preceding Example 4.14. The maximum angular differences are

$$|\theta_1 - \theta_2| = 0.049 \text{ rad}(2.81°)$$

and

$$|\theta_2 - \theta_3| = 0.013 \text{ rad}(0.745°)$$

The efficiency curve for this system is shown in Figure 4.30, where the peak efficiency value is approximately 87%.

Example 4.16

In Figure 4.31 the efficiency–wave number relation is presented for condition 3, for which the 450-ft (137-m) contouring raft system is exposed to a *double-peaked wave spectrum*. This spectrum, representative of that observed in the *North Sea*, is also shown in the figure. The two peaks respectively correspond to a persistent *swell component* (long-wave, low-wave-number value) and the *wind-generated sea* with its peak occurring at the higher wave number. Referring to the efficiency curves in Figure 4.31, it can be seen that the four-raft system has a well-defined response at both spectral peaks, whereas the response of the three-raft system is broadband with its maximum efficiency (~98%) in the neigh-

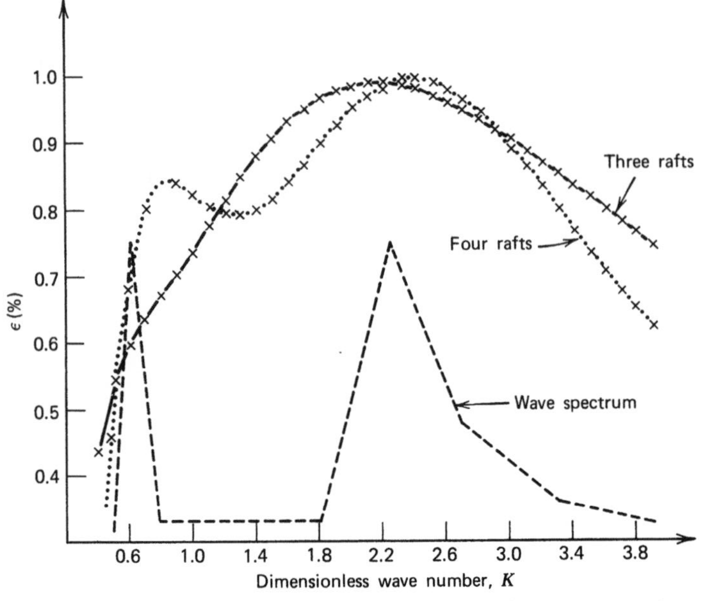

Figure 4.31 Raft efficiency as a function of wave number for three- and four-raft systems in a random sea. From Haren (1978). The raft system length is assumed.

borhood of the peak of the wind-generated wave spectrum. The curve for the three-raft system results from optimized design lengths of

$$L_1 = 19.6 \text{ ft } (5.97 \text{ m})$$

$$L_2 = 208 \text{ ft } (63.4 \text{ m})$$

$$L_3 = 222 \text{ ft } (67.7 \text{ m})$$

Example 4.17

Our last contouring raft example results from the optimization under condition 4. The optimum design lengths for this condition are

$$L_1 = 13.6 \text{ ft } (4.14 \text{ m})$$

$$L_2 = 76.4 \text{ ft } (23.3 \text{ m})$$

$$L_3 = 0.600 \text{ ft } (0.183 \text{ m})$$

with a system length of 90.6 ft (27.6 m). The efficiency curves for condition 4 are presented in Figure 4.32, where the peak efficiency value is about 86% for the three-raft configuration. We see that the response is significant for the wind-generated spectrum only.

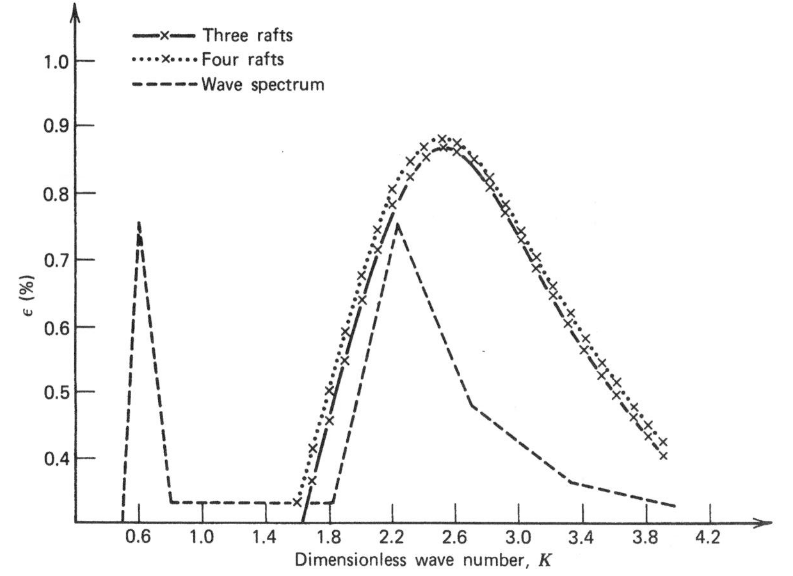

Figure 4.32 Raft efficiency as a function of wave number for three- and four-raft systems in a random sea. From Haren (1978).

If the results of Examples 4.14 through 4.17 are compared, condition 3 appears to be the best overall design because (*a*) the response is of broadband in a real sea spectrum having two distinct peaks and (*b*) the maximum efficiency is about 98% based on a two-dimensional wave condition. The primary drawback in this optimum design is due to its large size, which affects the cost of the system.

The efficiency curves of Haren (1978) presented in Figures 4.29 through 4.32 are based on two-dimensional waves where the crest length and contouring raft system width are equal. We observe peak efficiency values in all the curves for the three-raft configurations. Thus the peak value corresponds to a *resonance condition*. Based on the results in Section 4.2, for the nodding duck and on the results in Section 4.2, D for wave focusing, we can expect efficiencies greater than 100% when the contouring rafts are operating in three-dimensional waves because the interaction between the incident and radiant waves causes a *focusing* of wave energy on the raft system when the incident wave frequencies are in the neighborhood of the natural frequency of the system.

There are two *major problems* associated with contouring raft wave energy conversion. First, the *physical size* of the raft system is excessive. For example, using the results of condition 3 to efficiently ($\geqslant$ 80%) capture the power from waves having an average power per crest length of 5 kW/ft (16.4 kW/m) to supply a coastal town of 1000 homes with 5 kW per home, a raft system occupying a surface area of 5.62×10^5 ft^2 (5.23×10^4 m^2) or 12.9 acres would be required. The second major problem concerns the *mooring* of the rafts in

Plate 4.4 An artist's concept of a Cockerell raft wave energy conversion system. Courtesy of the Energy Technology Support Unit, AERE Harwell—The United Kingdom.

heavy seas. Results of the United Kingdom Wave Energy Conversion Symposium show that the mooring costs would make contouring raft wave energy conversion cost-ineffective.

C Russell's Rectifier

Robert Russell, director of the Hydraulics Research Station at Wallingford, Oxfordshire, England, has devised a wave energy conversion scheme designed to operate in the coastal zone. This system, which Russell calls the *HRS rectifier*, is a lock-type structure having wave-activated one-way gates facing the sea. There are two sets of these gates—one set allowing influx to a *supply reservoir* and the other allowing only the efflux from a catch basin. The two basins are connected by a hydraulic turbine that is constantly turning because of a *hydraulic head* maintained by the wave action. A sketch of the system is shown in Figure 4.33.

Referring to Figure 4.33, the gates leading to the reservoir are forced open by the wave-induced pressures. The reservoir fills, and the gates are then closed when the internal hydrostatic pressure force exceeds the wave-induced force. The captured water within the reservoir then drains through the hydraulic turbine into the catch basin. This basin fills until the wave-induced pressure force on the exhaust gates becomes less than the internal hydrostatic

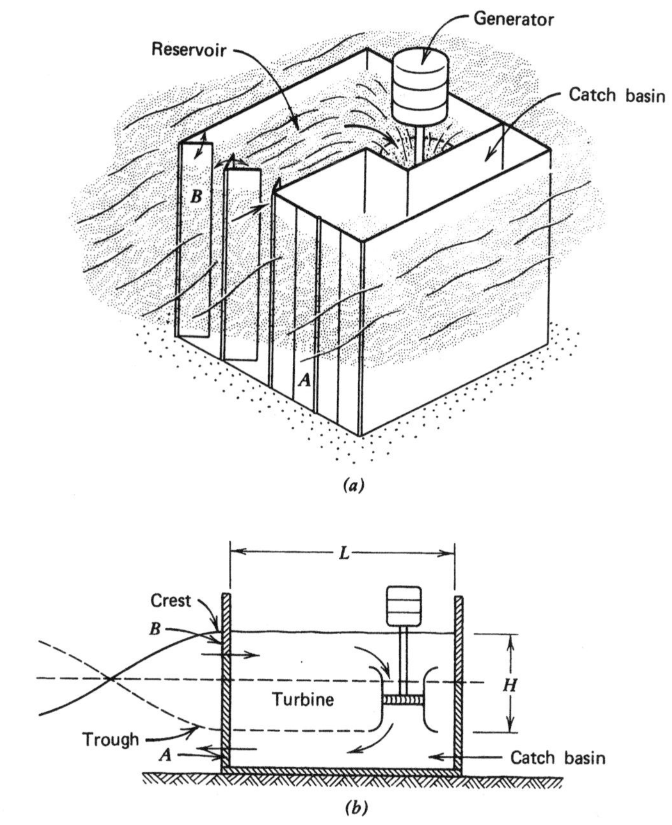

Figure 4.33 Sketch and diagram of the Russell rectifier: (*a*) artist's sketch of the Russell rectifier; (*b*) sketch and diagram of the Russell rectifier.

pressure force. The exhaust gates then open and the catch basin partially empties. These gates again close when the external free surface begins to rise, and the cycle is repeated. The cycle is depicted schematically in Figure 4.34.

The energy available to the turbine is the *potential energy* of the water column of the turbine passage. If the turbine flow area is A, the potential energy at any instant is

$$E_R^1 = \frac{\rho g \Delta_R^2 A}{2} \tag{4.103}$$

where Δ_R is the hydraulic head (height of the water column) defined by

$$\Delta_R = \delta_u(t) + \frac{H}{2} - \delta_l(t) \tag{4.104}$$

where $\delta_u(t)$ is the level of the reservoir above the SWL and $\delta_l(t)$ is the level of the catch basin above $z = -H/2$.

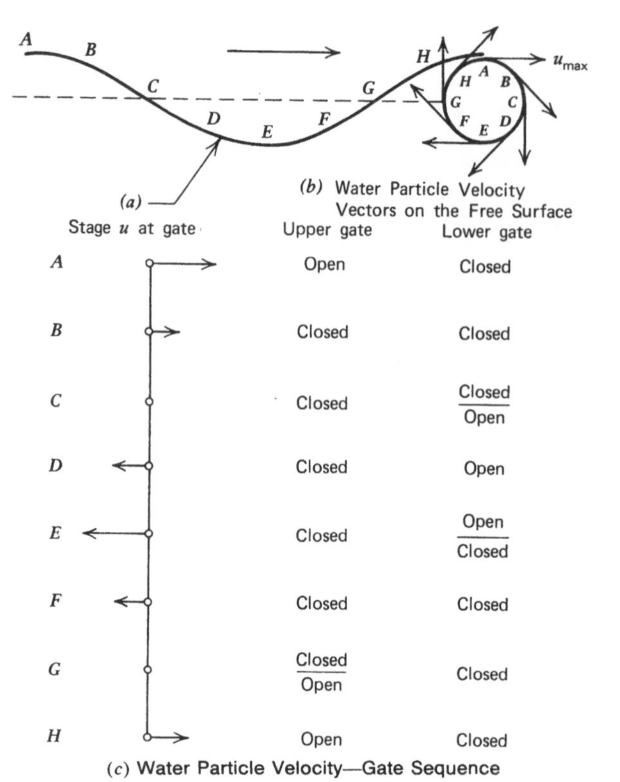

Figure 4.34 Water particle horizontal velocity component and gate sequences: (*a*) wave profile; (*b*) water particle velocity vectors on the free surface; (*c*) horizontal particle velocity–gate sequence.

To gain an understanding of the behavior of Δ_R over one wave period T, consider the idealized sequence shown in Figure 4.35. The reservoir is assumed to be "full" at time $t = 0$; that is, the basin level is equal to half the wave height or, mathematically, $\delta_u(0) = H/2$. Also, at $t = 0$ the lower basin level is assumed to be $\delta_l(0) = H/3$. Thus from equation (4.104) the *initial head* is $\Delta_R(0) = 2H/3$. The basin levels are assumed to change *linearly* with time when the gates are closed. This is not exactly the case; however, with proper design of the turbine passage, this linear variation can be well approximated. See the fluid mechanics text by Walther Kauffmann (1963) for an excellent discussion on time-dependent flows in passages. When $T/4 \leqslant t < T/2$, the wave-induced pressure force on the exhaust gates is less than internal hydrostatic pressure force within the catch basin; thus the gates open and the basin drains, with its level lowering as the external water level lowers. Over the time period $T/2 \leqslant t < 3T/4$ both sets of gates are closed and the reservoir continues its linear draining. Finally, when $3T/4 \leqslant t < T$, the wave pressure forces the upper gates open and the reservoir fills, and its level is described by $\delta_u(t) = \eta_0(t)$, where $\eta_0(t)$ is the *external water level* at the gates. We can

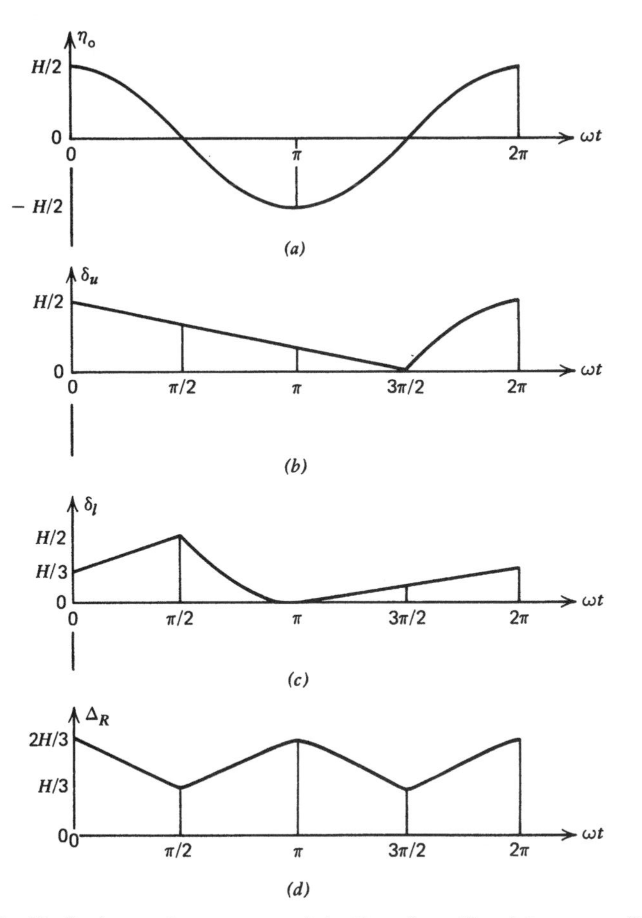

Figure 4.35 Idealized operating sequence of the Russell rectifier: (*a*) wave profile at the gates over one wave period; (*b*) water level in the reservoir over one wave period; (*c*) water level in the catch basin over one wave period; (*d*) operating head over one wave period.

summarize this sequence as follows:

$$\left(0 \leqslant t < \frac{T}{4}\right): \qquad \Delta_R = \frac{2H}{3}\left[1 - \frac{2t}{T}\right] \tag{4.105a}$$

$$\left(\frac{T}{4} \leqslant t < \frac{T}{2}\right): \qquad \Delta_R = \frac{H}{2}\left[1 - \frac{4}{3}\frac{t}{T} - \cos\left(\frac{2\pi t}{T}\right)\right] \tag{4.105b}$$

$$\left(\frac{T}{2} \leqslant t < \frac{3T}{4}\right): \qquad \Delta_R = \frac{4H}{3}\left[1 - \frac{t}{T}\right] \tag{4.105c}$$

$$\left(\frac{3T}{4} \leqslant t < T\right): \qquad \Delta_R = \frac{H}{2}\left[\frac{5}{3} - \frac{4t}{3T} + \cos\left(\frac{2\pi t}{T}\right)\right] \tag{4.105d}$$

Combining the results of equations (4.103) through (4.105) and integrating the result over one wave period, we obtain the *energy available to the turbine* over one wave period. This is

$$\begin{aligned}\hat{E}_R &= \frac{\rho g A}{T}\int_0^T \Delta_R^2\,dt \\ &= \frac{\rho g H^2 A}{T}\left\{\frac{4}{9}\int_0^{T/4}\left[1-\frac{2t}{T}\right]^2 dt\right. \\ &\quad + \frac{1}{4}\int_{T/4}^{T/2}\left[1-\frac{4t}{3T}-\cos\left(\frac{2\pi t}{T}\right)\right]^2 dt \\ &\quad + \frac{16}{9}\int_{T/2}^{3T/4}\left[1-\frac{t}{T}\right]^2 dt \\ &\quad \left.+ \frac{1}{4}\int_{3T/4}^{T}\left[\frac{5}{3}-\frac{4t}{3T}+\cos\left(\frac{2\pi t}{T}\right)\right]^2 dt\right\} \\ &= 0.123\rho g H^2 A \end{aligned} \tag{4.106}$$

From the results of equation (4.106) we see that the greater the flow area A, the greater the energy conversion. Naturally, there is a practical limit to the flow area, and that limit is determined by the width B of the catch basin. The maximum flow area value is then

$$A_{\max} = \frac{\pi B^2}{4} \tag{4.107}$$

To enhance the energy conversion ability of the rectifier, the length L of the catch basin should be such that *harbor resonance* occurs in a monochromatic sea. This phenomenon is well described in the book by Wiegel (1964). Furthermore, in a paper by Cranfield (1979), it is mentioned that the *optimum length* is

$$L_{\text{opt}} = \frac{\lambda}{5} \tag{4.108}$$

which is an indication of harbor resonance. When resonance occurs, the height of the water within the basin is considerably greater than the wave height H. For an ideal *standing wave* in the basin, the height H_{opt} is twice that of the wave; thus

$$H_{\text{opt}} = 2H \tag{4.109}$$

Assuming an ideal design of the rectifier, the efficiency is

$$\begin{aligned}\epsilon &= \frac{\hat{E}_R}{E}\\ &= \frac{0.123\rho g H_{opt}^2 A_{max}}{\rho g H^2 \lambda B/8}\\ &= 3.09\frac{B}{\lambda}\end{aligned} \tag{4.110}$$

where the results of equations (2.17) and (4.106) through (4.109) are combined.

Example 4.18

In the article by Cranfield (1979) a rectifier is sketched operating in shallow water. The capture width B of the three-cell system (three catch basin–exhaust basin pairs) is 98.4 ft (30 m), whereas the length L of the basins is 98.4 ft (30 m). Although the flow area A of the rectifier is not given, we assume a maximum flow area for each basin as given by equation (4.107). Since there are three cells, the flow area is

$$A_{max} = 3\frac{\pi B_{cell}^2}{4} = 2530 \text{ ft}^2 \ (235 \text{ m}^2)$$

where $B_{cell} = B/3 = 32.8$ ft (10 m). Assume a wave height H of 16.4 ft (5 m) and a water depth d of 57.4 ft (17.5 m). Harbor resonance is assumed such that equations (4.108) and (4.109) apply. Thus the design wavelength λ and the basin wave height H_{opt} are 492 ft (150 m) and 32.8 ft (10 m), respectively. The energy available to the turbines, using the results of equation (4.106), is

$$\begin{aligned}\hat{E}_R &= 0.123\rho g H_{opt}^2 A_{max}\\ &= 0.123(2.00)32.2(32.8)^2(2530)\\ &= 2.16\times 10^7 \text{ lb-ft } (2.92\times 10^7 \text{ N-m})\end{aligned}$$

The ideal efficiency of the three-cell system is obtained from equation (4.110) where $A_{max} = 3(\pi B_{cell}^2/4)$; thus

$$\begin{aligned}\epsilon &= \frac{0.123\rho g(4H^2)3(\pi B_{cell}^2/4)}{\rho g H^2 \lambda 3 B_{cell}/8}\\ &= \frac{3.09 B_{cell}}{\lambda}\\ &= \frac{3.09(32.8)}{492}\\ &= 0.206 \quad \text{or} \quad 20.6\%\end{aligned}$$

The results in Example 4.18 show that an energy conversion efficiency of 20.6% can be obtained for a rectifier operating under ideal conditions in a monochromatic sea. The ideal conditions include a harbor resonance condition where the basin wave height is twice that of the incident wave height. Under less ideal conditions, including *real seas* or random waves, the expected efficiencies would be somewhat less. Considering the *capital costs* involved, the rectifier does not appear to be a strong candidate wave energy conversion system.

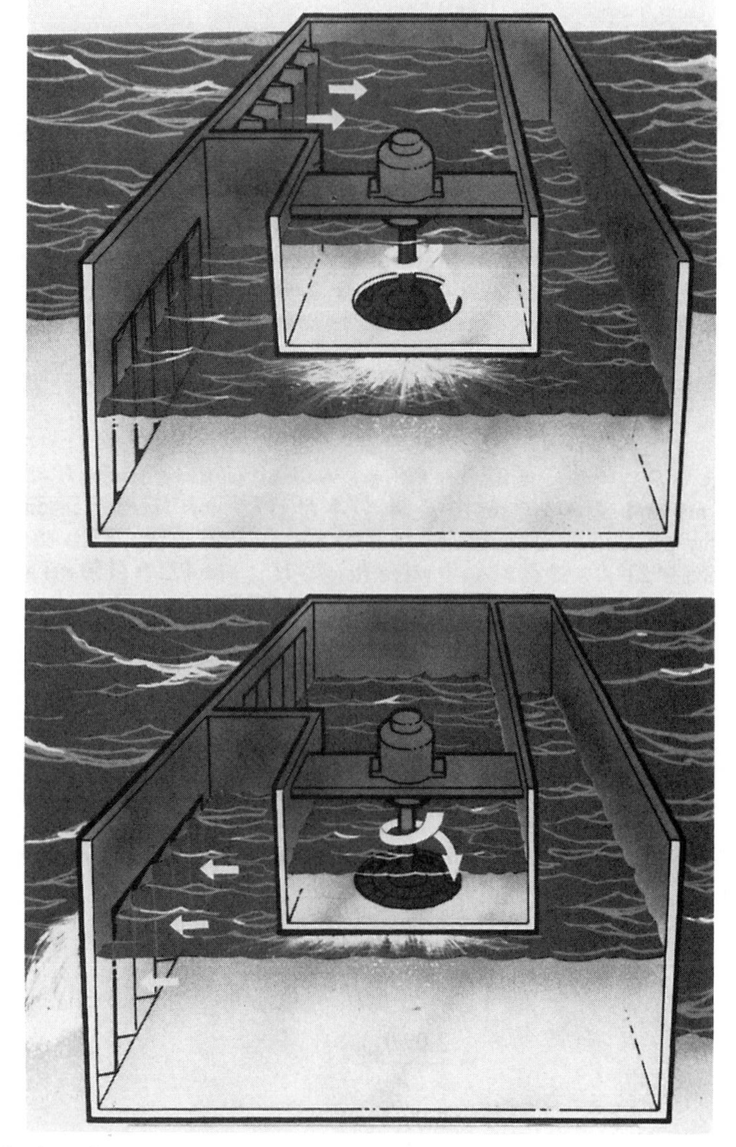

Plate 4.5 An artist's concept of the operation of the Russell rectifier. Courtesy of the Energy Technology Support Unit, AERE Harwell—The United Kingdom.

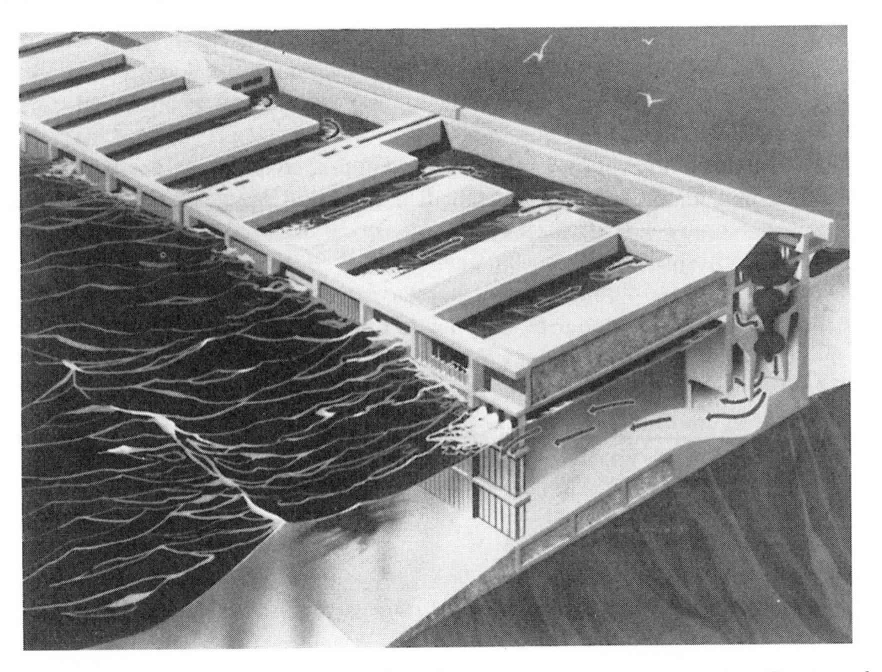

Plate 4.6 An artist's concept of a Russell rectifier wave energy conversion system. Courtesy of the Energy Technology Support Unit, AERE Harwell—The United Kingdom.

D Wave Focusing Techniques

With the realization that it is impractical to have large sections of coastal waters occupied by wave energy conversion devices, engineers began to seek methods of *focusing* wave power on conversion devices occupying relatively small regions. Their efforts have resulted in three very promising techniques: the "antenna" effect caused by interference of radiant waves from resonant body motions and incident waves, "lens" focusing due to refraction over a lens-shaped submerged platform, and "island" focusing caused by refraction on the slopes of an artificial "atoll." These techniques make it possible to have effective wave energy conversion with a minimum *navigation blockage* and few *environmental problems.* In this section we outline the basic theoretical aspects of these three wave focusing techniques.

The Antenna Effect

As the term "antenna effect" implies, this water wave focusing phenomenon is similar to that of radiowaves incident on a tuned antenna. The primary source of information concerning this technique is the University of Trondheim in Norway. At the Institute of Experimental Physics of that University Johannes Falnes and Kjell Budal have investigated the antennalike focusing of water waves—both theoretically and experimentally. Their efforts and those of their colleagues have resulted in the publications of Budal and Falnes (1975a, 1975b, 1977, 1979), Budal (1977), and Ambli et al. (1977). The basis for their efforts is the theoretical work of Newman (1962) describing the forces on symmetric bodies in waves. Newman has also extended his work in this area, as evidenced by his 1976 publication, among others. Evans (1976) has also theoretically studied the phenomenon.

An excellent description of the phenomenon is given by Falnes and Budal (1978). Consider the sequence sketched in Figure 4.36. A right-running incident wave (Figure 4.36*a*) excites a heaving body that, in turn, radiates waves of the same frequency as the incident wave if the body is in resonance, as illustrated in Figure 4.36*b*. Ideally, the heaving motions can extract only half of the wave energy; therefore, the heights of the waves radiating from the heaving body are much less than that of the incident waves. The combination of the incident and radiant wave systems results in an increased wave height to the left of the body and a reduced wave height to the right (Figure 4.36*c*). For a body undergoing pure pitching (or rolling) motions, the radiant wave

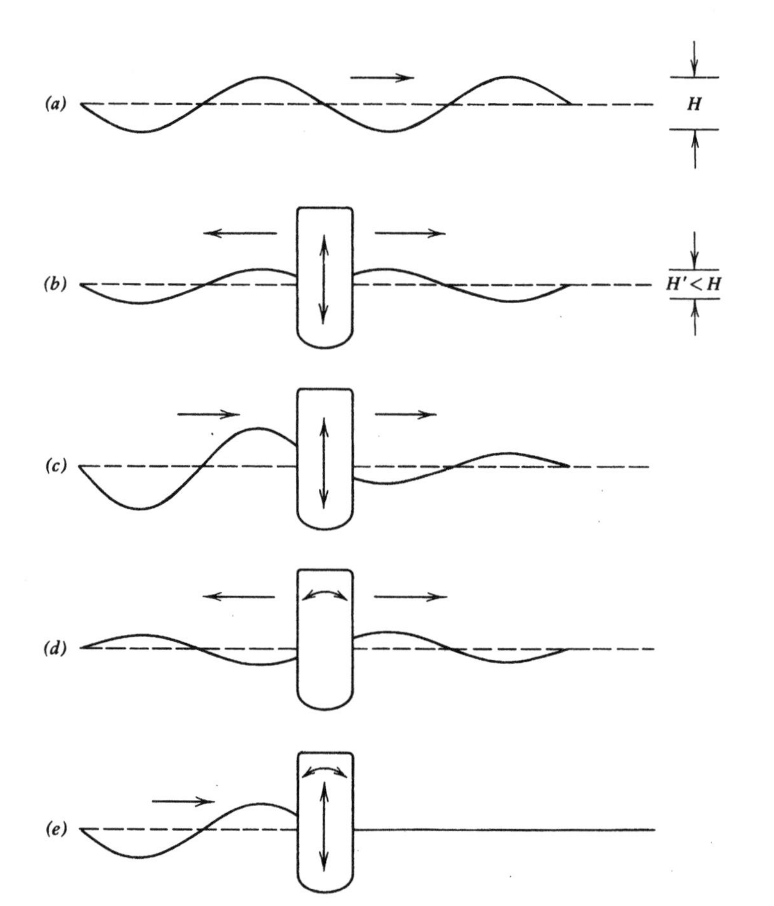

Figure 4.36 Illustration of two-dimensional interference: (*a*) right-running incident wave; (*b*) heave-induced radiant waves: (*c*) superposition of (*a*) and (*b*); (*d*) pitching-induced radiant waves; (*e*) superposition of (*a*), (*b*), and (*d*).

patterns are 180° out of phase with the incident waves (Figure 4.36*d*). Thus for a body undergoing both resonant heaving and pitching (or rolling), the theoretical result is total extraction of the incident wave energy as shown in Figure 4.36*e*.

Extending this reasoning to three dimensions, Falnes and Budal (1978) find that the interference of the radiant and incident waves cause the antennalike focusing of the incident wave energy on the resonating body. Consider the heaving body sketched in Figure 4.37*a*. Falnes and Budal (1978) predict that the *time-averaged wave power* absorbed by the body is

$$\hat{P}_0 = \tfrac{1}{2} F_0 V_0 \cos(\psi) - \tfrac{1}{2} R_0(\omega) V_0^2 \tag{4.111}$$

where F_0 is the vertical wave force amplitude, V_0 is the amplitude of the vertical velocity of the body, ψ is the phase angle between the body motion

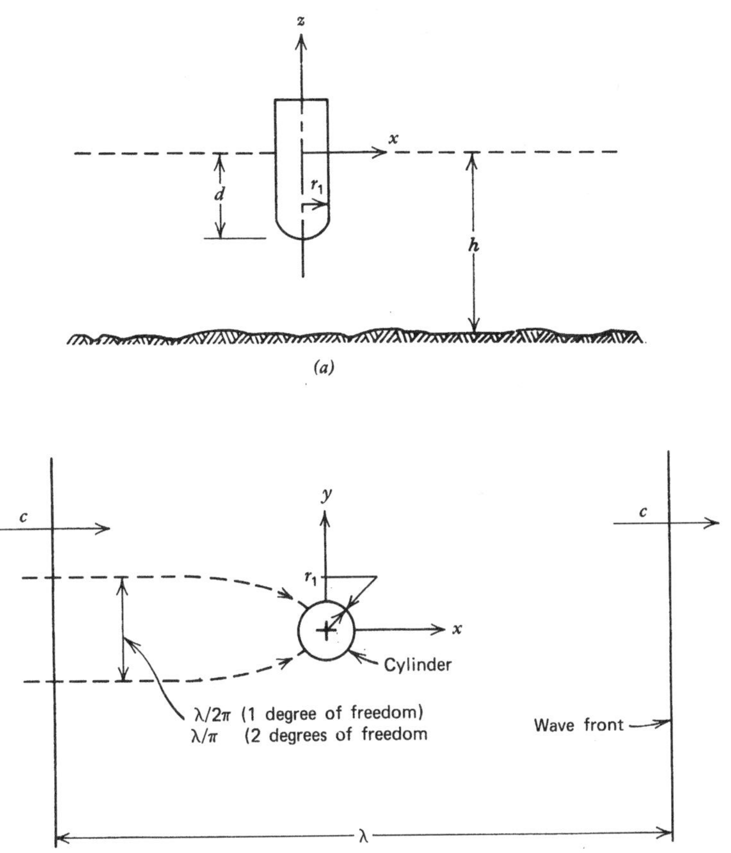

Figure 4.37 Conceptual sketch of "antenna" focusing: (*a*) notation; (*b*) areal sketch of focused wave power.

and the wave force and $R_0(\omega)$ is the *radiation resistance* that depends on the shape of the body and the motion frequency. Since the body is in resonance with the wave, the motion frequency and the wave frequency ω are the same. From equation (2.19), the wave power per unit crest length is

$$\frac{P}{b} = \frac{\rho g H^2 c_g}{8} \tag{4.112}$$

where b is an arbitrary crest length and c_g is the group velocity, which is defined by equation (2.20a) as

$$c_g = \frac{c}{2}\left[1 + \frac{2kh}{\sinh(2kh)}\right] \tag{2.20a}$$

Returning to equation (4.111), the *optimum power* absorbed by the body occurs when the wave force and body motions are *in phase*, that is, when $\psi = 0$, under the condition of

$$V_0 = \frac{F_0}{2R_0} \tag{4.113}$$

Thus the *optimum average absorbed power* is obtained from equation (4.111) using the results of equation (4.113).

$$\hat{P}_{\text{opt}} = \frac{F_0^2}{8R_0} \tag{4.114}$$

For a *heaving circular cylinder*, Newman (1962) shows that the wave force amplitude is

$$F_0 = \sqrt{\frac{\rho g H^2 c_g \lambda R_0}{2\pi}} \tag{4.115}$$

The combination of equations (4.114) and (4.115) yields

$$\hat{P}_{\text{opt}} = \frac{\rho g H^2 c_g \lambda}{16\pi}$$

$$= \frac{P\lambda}{2\pi b} \tag{4.116}$$

where P is the wave power given in equation (4.112). If we replace the arbitrary crest length b by the diameter of the circular cylinder $2r_1$ in equation (4.116), then that equation states that *the optimum average power available to the circular cylinder is equal to the wave power within a crest length of* $\lambda/2\pi$. Since $\lambda/2\pi$ can be many times that of the cylinder's diameter, we see that the wave power is, in fact, *focused* on the body undergoing the *single degree of freedom motion* as illustrated in Figure 4.37*b*.

For a body undergoing resonant motions in *two degrees of freedom*—for example, heaving and pitching—the optimum averaged absorbed power is

$$\hat{P}_{\text{opt}} = \frac{P\lambda}{\pi b} \tag{4.117}$$

Thus in this situation the *wave power is focused from a crest length of* λ/π, that is twice the length of that for the single degree of freedom (see Figure 4.37*b*).

Example 4.19

Consider the water column of a *stationary pneumatic wave energy convertor* in Figure 4.9 to be the body in motion. When the wave frequency equals the cavity resonant frequency f_c in equation (4.35), the water experiences maximum vertical motions. The diameter of the water column D_1 is 3 ft (0.914 m), and the cavity resonant period T_c is 3.50 secs. The device is the same as that described in Example 4.4. For purposes of illustration, assume that the wave energy conversion is in *deep water* so that the wavelength corresponding to the cavity resonance period is, according to equation (2.6),

$$\lambda = \frac{gT^2}{2\pi}$$

$$= \frac{32.2(3.50)^2}{2\pi}$$

$$= 62.6 \text{ ft } (19.1 \text{ m})$$

Thus the energy conversion system is situated in water of a depth $h \geqslant \lambda/2 = 31.3$ ft (9.55 m). Since this is a *single-degree-of-freedom system* the optimum power available to the device is given by equation (4.116). Of more interest here, however, is the *efficiency* of the system based on the available wave power in a crest length equal to the water column diameter D_1. This is found by rearranging equation (4.116) to obtain

$$\frac{\hat{P}_{\text{opt}}}{P} = \frac{\lambda}{2\pi D_1}$$

$$= \frac{62.6}{2\pi(3)}$$

$$= 3.33 \quad \text{or} \quad 333\% \tag{4.118}$$

This result shows that the power is *focused* on the device from a crest length of $3.33D_1$ or 9.99 ft (3.04 m). If the pneumatic wave energy convertor itself is 33.3% efficient, the overall efficiency of the conversion technique (including focusing) is

$$\epsilon\left(\frac{\hat{P}_{opt}}{P}\right) = \frac{\epsilon\lambda}{2\pi D_1}$$

$$= 0.333(3.33)$$

$$= 1.11 \quad \text{or} \quad 111\% \tag{4.119}$$

Now consider a *row* of heaving bodies each having symmetry about their vertical axes, as shown in the areal view sketched in Figure 4.38. The waves approach the row such that the wave front is at an angle β to the row. Budal (1977) shows that the *radiation resistance* on any symmetric body within the row (excluding those on the ends of the row) is given by

$$R = \frac{\lambda R_0}{\pi l \cos(\beta)} \tag{4.120}$$

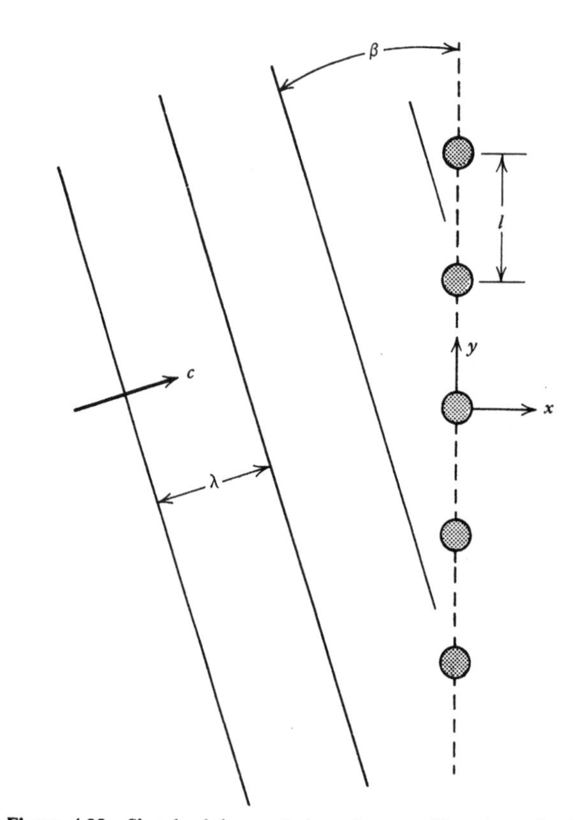

Figure 4.38 Sketch of the areal view of a row of heaving cylinders.

where R_0 is the radiation resistance of an isolated body and l is the separation distance of the bodies. The *wave force* in this case is

$$F_0 = \sqrt{\tfrac{1}{2}\rho g H^2 c_g R l \cos(\beta)} \tag{4.121}$$

The *optimum average power available* to each heaving body, from equation (4.114), is

$$\begin{aligned}\hat{P}_{opt} &= \frac{F_0^2}{8R} \\ &= \frac{\rho g H^2 c_g l \cos(\beta)}{16} \\ &= \frac{Pl\cos(\beta)}{2b} \end{aligned} \tag{4.122}$$

where the results of equation (4.112) are used. From the results of equation (4.122), we see that the *maximum power* available to each body within the row corresponds to the condition of $\beta = 0$, that is, when the wave front is aligned with the row.

Example 4.20

Consider a row of stationary pneumatic wave energy convertors (Figure 4.9) of the type described in Example 4.19, that is, having a water column diameter D_1 of 3 ft (0.914 m) and a resonant period T_c of 3.50 sec. Waves approach the row at an angle β, as illustrated in Figure 4.38. For any one of the conversion devices within the row, the optimum wave power available is obtained from equation (4.122), where b in that equation is replaced by D_1; thus

$$\hat{P}_{opt} = \frac{Pl\cos(\beta)}{2D_1}$$

Comparing this expression with that of equation (4.116), we see that the *upper limit on the separation length* l is

$$l = \frac{\lambda}{\pi\cos(\beta)} \tag{4.123}$$

Thus equation (4.122) applies when $l < \lambda/\pi\cos(\beta)$. Otherwise, for a single-degree-of-freedom system, equation (4.116) yields the optimum power.

Assume a deep water sinusoidal wave condition and a separation length l of 18 ft (5.49 m). The value of l, then, is approximately $\lambda/3.5$ in this case. If the wave front is aligned with the row, that is, $\beta = 0$, the optimum power available to each device is

$$\hat{P}_{\mathrm{opt}} = \frac{Pl\cos(0)}{2D_1}$$

$$= \frac{P(18)1}{2(3)}$$

$$= 3.0P$$

where P is the wave power of a crest length equal to the water column diameter D_1. If each device in the row is 33.3% efficient, the overall efficiency is

$$\epsilon\left(\frac{\hat{P}_{\mathrm{opt}}}{P}\right)\Bigg|_{\beta=0^0} = 0.333(3.0)$$

$$= 0.999 \quad \text{or} \quad 99.9\%$$

If the waves approach the row at $\beta = 30^0$, the overall efficiency is

$$\epsilon\left(\frac{\hat{P}_{\mathrm{opt}}}{P}\right) = \frac{\epsilon l\cos(30^0)}{2D_1}$$

$$= 0.333(3.0)(0.866)$$

$$= 0.865 \quad \text{or} \quad 86.5\%$$

Some *caution* must be taken in the interpretation of the results of this section. As Falnes and Budal (1978) state, the optimum conditions are rather specific. For a heaving type of motion, the wave force and body motions must be in resonance to achieve the results of equations (4.116) and (4.122) for a sinusoidal wave. The phase angle can be controlled by adjusting the mass of a heaving body by using a *variable ballast* system or by altering the draft (water column length) of a pneumatic system. *Confidence* in the focusing phenomenon, however, is obtained from the experimental results of Carmichael (1978) in Figures 4.25 and 4.26, which show overall efficiencies for a nodding duck model in excess of 100% over a rather broad wave frequency range.

Refraction Focusing—"DAM–ATOLL"

In 1946 Robert S. Arthur presented a paper in which he described the *refraction* of water waves by *idealized islands having circular bottom contours.* By the proper choice of bottom profile, Arthur (1946) showed that the power from that portion of the wave front directly influenced by the island is focused on the center of the island, as illustrated in Figure 4.39*a*. He studied the

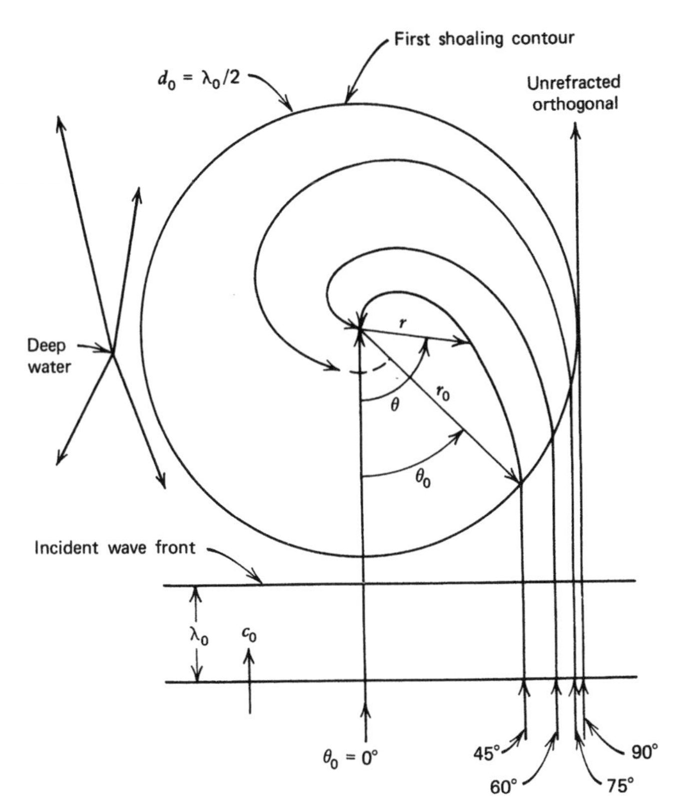

Figure 4.39a Schematic diagram of the top view of wave refraction by an ideal circular island.

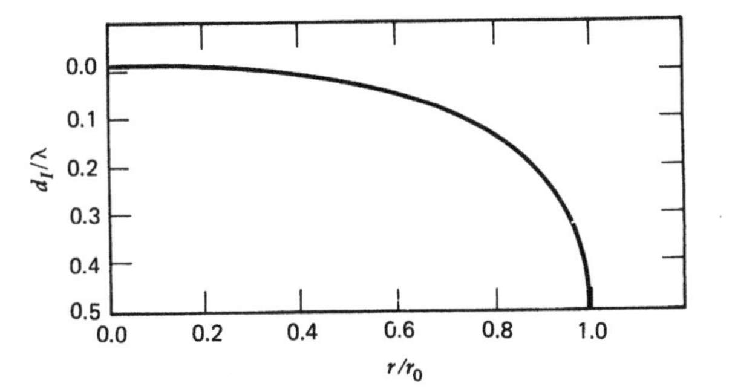

Figure 4.39b Dimensionless profile of an idealized island as given by equation (4.125).

focusing phenomenon by analyzing the changes in the directions of the wave *orthogonal* or *rays* (lines drawn normal to the wave front) as the wave refracts on the island.

Using the results of the Arthur (1946) study, Wirt (1976) of the Lockheed California Company showed that the most suitable island profile for wave energy conversion is the one that yields *logarithmic spiral orthogonals*, that is, those described by

$$r = r_0 \exp\left[(\theta - \theta_0)\cot(\theta_0)\right] \tag{4.124}$$

where r_0 is the radius of the first shoaling contour and θ_0 the angle of the nonrefracting orthogonal. Referring to Figure 4.39*b*, the profile of the island is given by the equation

$$d_I = \frac{\lambda_0}{2\pi}\left(\frac{r}{r_0}\right)\tanh^{-1}\left(\frac{r}{r_0}\right) \tag{4.125}$$

where λ_0 is the wavelength of the deep water wave. The waves refract in the pattern shown in Figure 4.39*a* until they *break*. The broken waves then become *surges* and are guided into a vertical shaft by vanes, as illustrated in Figure 4.40. The entrances of the guide vanes are positioned just before the breaking point of the waves. The vanes are designed to give the surge a tangential velocity as the water enters the vertical shaft, as illustrated in Figure 4.40. The water then descends as a water column with a rotational motion and thus acts as a *fluid flywheel*. At the bottom of the shaft is a vertical axis *turbine*, which is illustrated in the cross-sectional sketch in Figure 4.41. The turbine, in turn, drives an electrical generator. After exciting this turbine, the water then passes through a *radial diffuser* and returns to the ocean.

The system described by Wirt (1976), called the "DAM–ATOLL," is designed to adjust its vertical position as the mean water level changes with the tides. This is accomplished by giving the system a relatively small amount

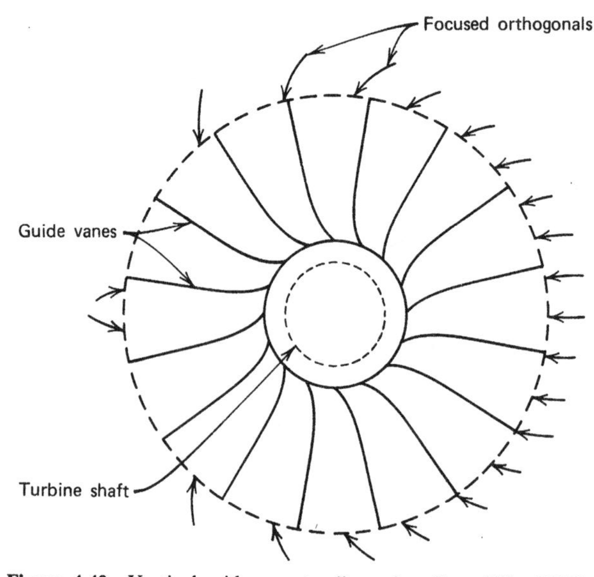

Figure 4.40 Vertical guide vane configuration. From Wirt (1976).

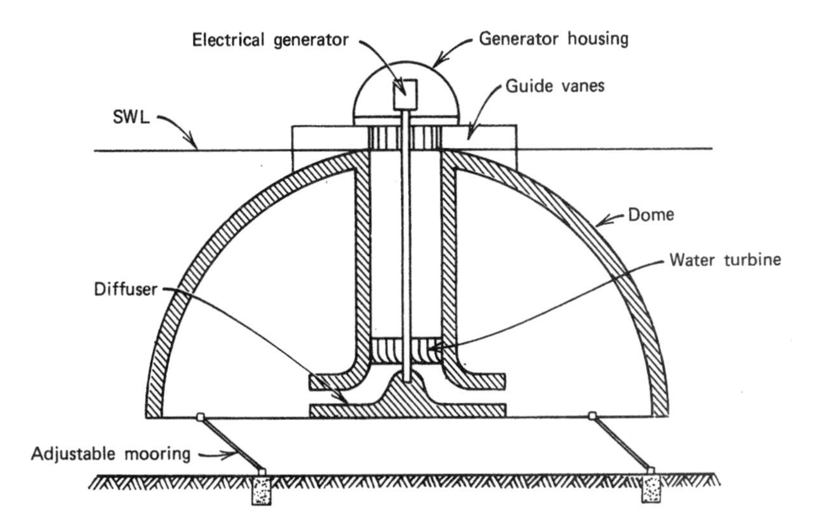

Figure 4.41 Schematic drawing of the vertical cross section of DAM-ATOLL. From Wirt (1976).

of positive buoyancy and by using an adjustable mooring, as illustrated in Figure 4.41. The island, or atoll, can be compliant walled and water filled. This design feature allows for ease in both transportation to the site and positioning. After the system is placed in the water, it floats as a result of flotation chambers located in the rigid central portion of the system. The flexible chambered walls of the atoll are then filled with water that is pumped in at a high pressure to maintain the desired profile [given by equation (4.125)]. These design features should keep the costs of the DAM–ATOLL system to a minimum. A sketch of an operational system is presented in Figure 4.42.

Example 4.21

Assume that the wave conditions of Examples 4.19 and 4.20 are experienced by DAM–ATOLL, that is, $H_0 = 3$ ft (0.914 m) and $T = 3.50$ sec. Furthermore, the depth of the water surrounding the wave energy converter is greater than half of the length of the incident waves; thus

$$h > \frac{\lambda_0}{2} = 31.3 \text{ ft } (9.55 \text{ m})$$

Therefore, the incident waves are deep water waves. The atoll is assumed to have a diameter of 200 ft (61.0 m) at its first refracting contour; thus

$$r_0 = 200 \text{ ft } (61.0 \text{ m})$$

$$d_0 = 31.3 \text{ ft } (9.55 \text{ m})$$

Figure 4.42 Artist's sketch of an operational DAM-ATOLL. Courtesy of the Lockheed-California Company.

The wave power available to the DAM–ATOLL is obtained from equation (4.112), where b in that equation replaced by $2r_0$. Thus in *deep water*, where the group velocity is

$$\begin{aligned} c_{g_0} &= \frac{c_0}{2} \\ &= \frac{\lambda_0}{2T} \\ &= 8.97 \text{ fps } (2.73 \text{ mps}) \end{aligned}$$

(where fps = feet per second and mps-meters per second) the *wave power* is

$$\begin{aligned} P &= \frac{\rho g H_0^2 c_{g_0} r_0}{4} \\ &= \frac{2.00(32.2)3^2(8.97)100}{4} \\ &= 130{,}000 \text{ lb-ft/sec } (176 \text{ kW}) \end{aligned}$$

If the power conversion *efficiency* ϵ is 33.3%, the electrical power delivered by the system is

$$\epsilon P = 0.333\,(130{,}000)$$

$$= 43{,}000 \text{ lb-ft/sec } (58.7 \text{ kW})$$

More recent information concerning the DAM–ATOLL concept can be found in the paper by Wirt and Higgins (1979). In addition, the scattering phenomenon by idealized islands has recently been studied by Skongaard et al. (1975).

"Lens" Focusing

In the preceeding subsection it was shown that the phenomenon of refraction on circular bottom contours of idealized islands results in the focusing of wave energy on the center of the islands. Now, we direct our attention to *focusing* by *refraction* on *submerged structures* having horizontal areas in the shape of *optical lenses.* An understanding of this focusing phenomenon can be obtained by considering the lens-shaped structure sketched in Figure 4.43. This configuration is discussed by McCormick in his 1979 and 1980 papers. Assume that the incident wave front is aligned with the front face of the lens structure and that the wave is in *deep water* ($h > \lambda_0/2$). Thus the wavelength and the phase velocity of this wave are, respectively,

$$\lambda_0 = \frac{gT^2}{2\pi} \tag{4.126}$$

and

$$c_0 = \frac{\lambda_0}{T} = \frac{gT}{2\pi} \tag{4.127}$$

The wave properties also exist in the lee of the structure, as shown in Figure 4.43*b*. If the depth d of the structure's platform is less than the half of the incident wavelength (i.e., $d < \lambda_0/2$), then the wavelength λ and the phase velocity c are both *decreased* over the structure, according to equations (2.3) and (2.4), respectively. The wave front, however, remains parallel with the incident wave fronts.

The lee edge of the platform is *circular*, having a radius of R'. As the wave passes over this edge the *orthogonal* or *wave ray* is turned by an angle

$$\alpha_L = \beta_0 - \beta_L \tag{4.128}$$

as illustrated in Figure 4.44. The angle β_0 is obtained from *Snell's law*, as discussed in Section 3.1:

$$\beta_0 = \sin^{-1}\left[\frac{c_0}{c}\sin(\beta_L)\right] \tag{4.129}$$

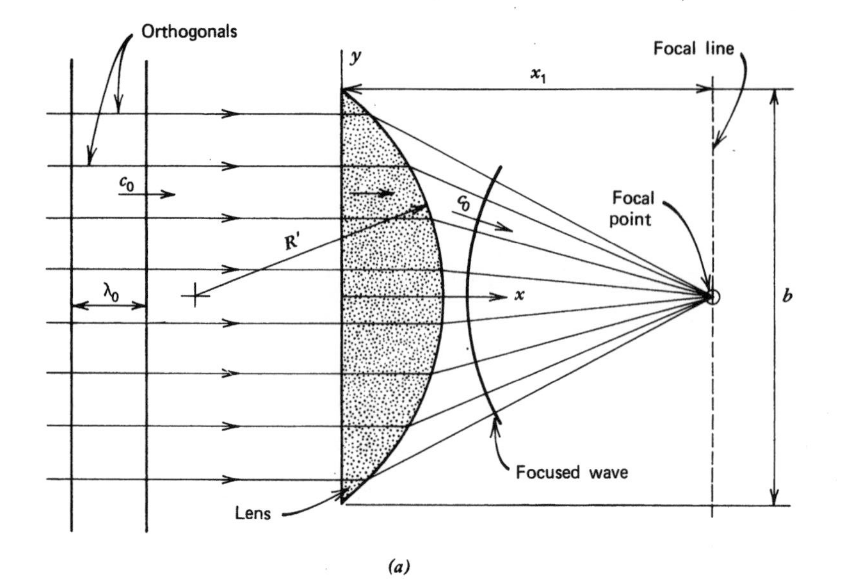

(a)

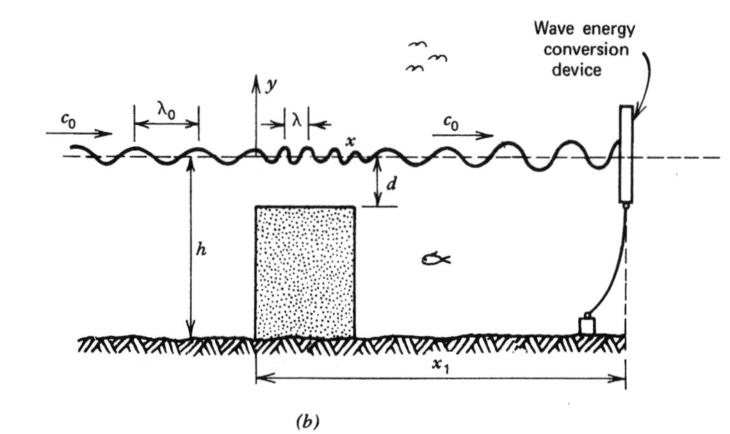

(b)

Figure 4.43 Sketches of the areal and side views of a lens-type wave focusing structure: (*a*) areal view of lens structure; (*b*) side view of lens structure showing cross section.

The rays are now focused on a point located a distance X_1 from the front face of the structure. This focal distance is obtained from the "*lens maker's equation*" (Halliday and Resnick, 1978),

$$\frac{1}{X_1} = \left(\frac{c_0}{c} - 1\right)\frac{1}{R'} \tag{4.128}$$

Thus,

$$X_1 = \frac{R'}{(c_0/c) - 1} \tag{4.129}$$

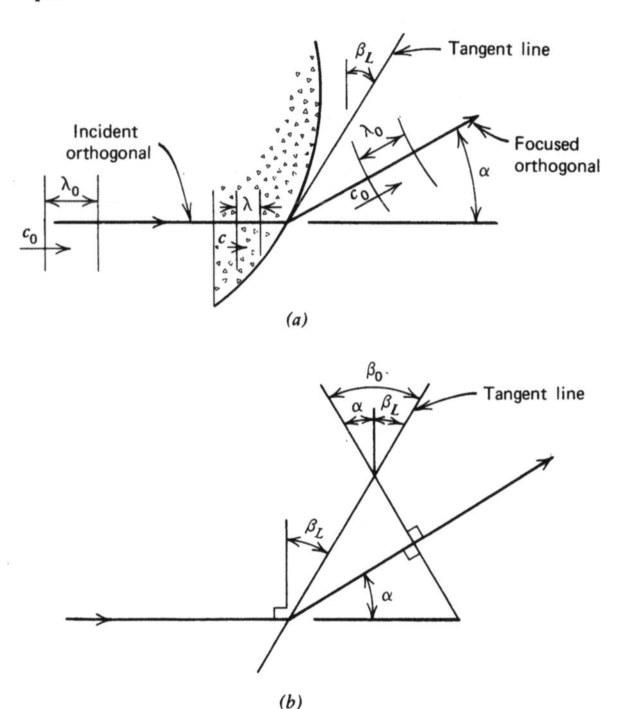

Figure 4.44 Notation for the application of Snell's law to lens focusing: (*a*) schematic diagram; (*b*) notation.

The *focal length can be decreased* by giving the front face of the structure a circular curvature with a radius R'', as illustrated in Figure 4.45. In this case the focal length is

$$X_2 = \frac{R'R''/(R' + R'')}{(c_0/c) - 1} < X_1 \tag{4.130}$$

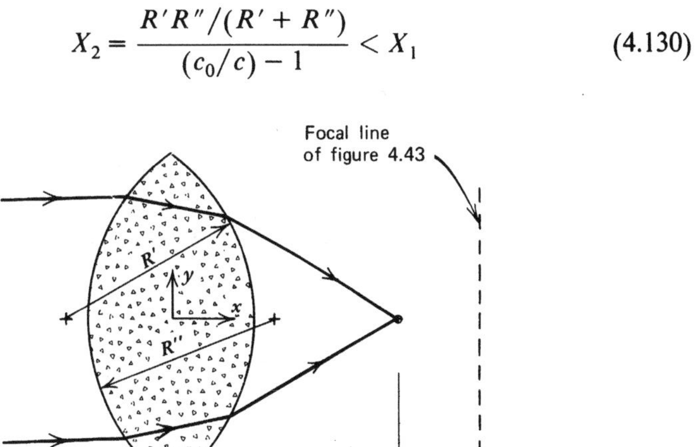

Figure 4.45 Areal view of a double-curvature lens structure.

***Note*:** When radii R' and R'' are equal, the focal length is

$$X_2 = \frac{R'/2}{(c_0/c) - 1} = \frac{X_1}{2} \tag{4.131}$$

Example 4.22

The DAM–ATOLL in Example 4.21 has a diameter of 200 ft (61.0 m) and experiences waves having a height H of 3 ft (0.914 m) and a period T of 3.50 sec. Consider a lens structure under the same wave conditions having a breadth b of 200 ft (61.0 m) and a platform depth d of 7.85 ft (2.29 m). Thus $d = \lambda_0/8$. From the results in Figure 2.2, (where d and h are the same), the ratio of the phase velocities for $d_0/\lambda_0 = 0.125$ is $c_0/c = 1.28$. The radii of curvature of both the front and lee faces of the structure are

$$R' = R'' = 200 \text{ ft } (61.0 \text{ m})$$

Thus, using the results of equation (4.131), the focal length is

$$X_2 = \frac{R'/2}{(c_0/c) - 1} = \frac{200/2}{1.28 - 1} = 357 \text{ ft } (109 \text{ m})$$

A wave energy conversion device located at the focal point is then exposed to a significant portion of the wave power of the 200 ft (61.0 m) of incident wave front.

***Note*:** A portion of the incident wave power will be reflected by the front face of the structure and, therefore, is unavailable to the conversion device. This *reflection* can be reduced by placing a *ramp* before the face, according to the results in Section 3.2.

When the configurations shown in Figures 4.43 and 4.45 are exposed to waves traveling in directions other than the x-direction, the focal point shifts in direction parallel to the y-axis, as illustrated in Figure 4.46. Thus, located at a distance X_2 from the y-axis, there is a *focal line*. To keep the wave energy conversion device at the focal point, a *station keeping* system must be used that simply consists of one or two small propulsion devices. These devices would be controlled by a servomechanism connected to wave direction sensor on the face of the lens structure.

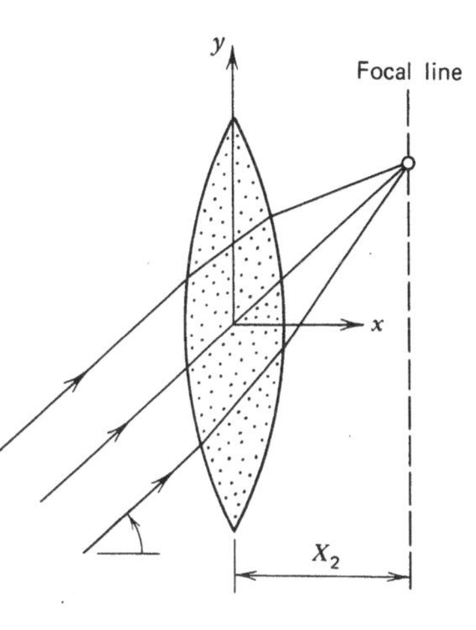

Figure 4.46 Waves approaching a double-curvature lens structure at an angle.

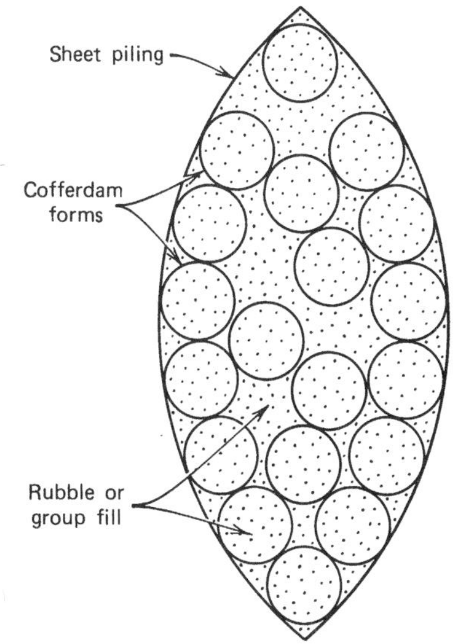

Figure 4.47 Possible structure of an ocean wave lens.

The use of lens focusing for wave energy conversion has been studied at the Central Institute for Industrial Research in Oslo, Norway, by Mehlum and Stamnes (1979) and at the U.S. Naval Academy by McCormick et al. (1980).

The construction of lens structures can be accomplished by using cofferdam structures filled with either stone or grout. The circular cylindrical shells, constructed of reinforced concrete, are lowered in place and then faced with sheet piling, as illustrated in Figure 4.47.

4.3 Summary

The nine basic wave energy conversion categories discussed in this chapter include many variations. The most basic examples in each category are presented to give the reader both a basic understanding of the operation and an expected efficiency range. A discussion of the systems aspects of these nine catagories can be found in the paper by Rogalski et al. (1979) in which one extremely important point is made concerning the overall performance of wave energy conversion systems, namely, that a distinction must be made between wave *energy extraction* and *energy conversion*. The former refers to the changing of wave energy into mechanical motion, and the latter concerns the conversion of that mechanical motion into usable energy. For example, Salter's nodding ducks (section 4.2, A) is one of the most efficient energy extractors, but also one of the most difficult in energy conversion.

Also discussed by Rogalski et al. (1979) is the fact that the overall assessment of each wave energy conversion system must include consideration of *construction*, *positioning*, *operation*, *maintenance*, and *survivability*. The oceans are the most hostile of environments. As we have learned in our experiences with offshore oil exploration and production, working in the ocean environment is both dangerous and expensive. Thus it might be more *cost effective* to sacrifice some overall wave energy conversion efficiency to ensure the safety of the system and, thereby, increase its lifetime.

References

Ambli, N., Budal, K., Falnes, J., and Sorenssen, A. (1977), "Wave Power Conversion by a Row of Optimally Operated Buoys," *Proceedings, 10th World Energy Conference*, Istanbul, September, Paper 4.5-2.

Arthur, R. S. (1946), "Refraction of Water Waves by Islands and Shoals with Circular Bottom Contours," *Transactfons, American Geophysical Union*, Vol. 27, No. 2, pp. 168–177.

Baird, W. F. (1968), "On Means of Utilizing the Energy of Wind Waves," M. S. thesis, Civil Engineering Department, Queen's University, Canada.

Bhattacharyya, R. (1978), *Dynamics of Marine Vehicles*, Wiley-Interscience, New York.

Budal, K. (1977), "Theory of Absorption of Wave Power by A System of Interacting Bodies," *Journal of Ship Research* (SNAME), Vol. 21, No. 4, pp. 248–253.

Budal, K., and Falnes, J. (1975a), "A Resonant Point Absorber of Ocean-Wave Power," *Nature*, Vol. 256, pp. 478–479.

Budal, K., and Falnes, J. (1956), "Power Generation from Ocean Waves Using a Resonant Oscillating System," *Marine Science Communications*, Vol. 1, pp. 269–288.

Budal, K., and Falnes, J. (1977), "Optimum Operation of Improved Wave-Power Converter," *Marine Science Communication*, Vol. 3, No. 2, pp. 133–150.

Budal, K., and Falnes, J. (1979), "Interacting Point Absorbers with Controlled Motion," *Proceedings, Power from Sea Waves Conference*, University of Edinburgh, Scotland, June.

Carmichael, A. D. (1978), "An Experimental Study and Engineering Evaluation of the Salter Cam Wave Energy Converter," Massachusetts Institute of Technology, Cambridge, Mass., Report No. MITSG 72-22, December.

Cranfield, J. (1979), "Interest in Wave Power Growing," *Ocean Industry*, February, pp. 67–76.

Evans, D. V. (1976), "A Theory for Wave-Power Absorption by Oscillating Bodies," *Journal of Fluid Mechanics*, Vol. 77, pp. 1–25.

Eshbach, O. W., Ed. (1975), *Handbook of Engineering Fundamentals*, 3rd ed., Wiley, New York.

Falnes, J., and Budal, K. (1978), "Wave-Power Conversion by Point Absorbers," *Norwegian Maritime Research*, Vol. 6, No. 4, pp. 2–11.

Halliday, D., and Resnick, R. (1978), *Physics for Students of Science and Engineering*, Wiley, New York.

Haren, P. (1978), "Optimal Design of Hagen-Cockerell Raft," M.S. thesis, Department of Civil Engineering, Massachusetts Institute of Technology, Cambridge, Mass.

Hooft, J. P. (1970), "Oscillatory Wave Force on Small Bodies," *International Shipbuilding Progress*, April.

Kaufmann, W. (1963), *Fluid Mechanics*, McGraw-Hill, New York.

Lamb, H. (1932), *Hydrodynamics*, 6th ed., Cambridge University Press, U.K.

Masuda, Y. (1971), "Wave Activated Generator," *Proceedings, International Colloquium on the Exposition of the Oceans*, Bordeaux, France, March.

Masuda, Y., and Miyazaki, T. (1979), "Wave Power Electric Generation Study in Japan," *Proceedings, Wave and Tidal Energy Symposium* (British Hydromechanics Research Associates), Paper C.

McCormick, M. E. (1973), *Ocean Engineering Wave Mechanics*, Wiley-Interscience, New York.

McCormick, M. E. (1974a), "A Parametric Study of a Wave-Energy Conversion Buoy," *Proceedings, Offshore Technology Conference*, May, Paper No. OTC 2125.

McCormick, M. E. (1974b), "An Analysis of a Wave-Energy Conversion Buoy," *Journal of Hydronautics* (*AIAA*), Vol. 8, No. 3, July, pp. 77–82.

McCormick, M. E. (1976), "A Modified Linear Analysis of a Wave-Energy Conversion Buoy," *Ocean Engineering*, Vol. 3, No. 3, pp. 133–144.

McCormick, M. E. (1979), "Ocean Wave Energy Conversion Concepts," *Proceedings, Oceans '79* (IEEE and MTS), San Diego, September, pp. 553–558.

McCormick, M., Carson, B., and Rau, D. (1975), "An Experimental Study of a Wave-Energy Conversion Buoy," *Marine Technology Society Journal*, Vol. 9, No. 3, pp. 39–42.

McCormick, M., Kastner, R., and Glover, L. (1980), "Water Wave Focusing by Submerged Lens-Shaped Structures," U.S. Naval Academy, Report EW-14-80, November.

Mehlum, E., and Stamnes, J. (1979), "On the Focusing of Ocean Swells and Its Significance in Power Production," preprint, Symposium on Wave Energy Utilization, Chalmers University, Gothenburg, Sweden.

Moody, G. W. (1979), "The NEL Oscillating Water Column: Recent Developments," preprint, Symposium on Wave Energy Utilization, Chalmers University, Gothenburg, Sweden.

Mynett, A., Serman, D., and Mei, C. (1979), "Characteristics of Salter's Cam for Extracting Energy from Ocean Waves," *Applied Ocean Research*, Vol. 1, No. 1, pp. 13–20.

Newman, J. N. (1962), "The Exciting Forces on Fixed Bodies in Waves," *Journal of Ship Research* (SNAME), Vol. 6, No. 3, pp. 10–17.

Newman, J. N. (1976), "The Interaction of Stationary Vessels with Regular Waves," *Proceedings, 10th Symposium of Naval Hydrodynamics* (U.S. Office of Naval Research), London.

Palme, A. (1920), "Wave Motion Turbine," *Power*, Vol. 52, No. 18, November, pp. 200–201.

Rogalski, W., Midboe, E., Sherwood, W., and Szeto, F. (1979), "The State of the Art in Alternate Ocean Energy Systems," meeting of Cheasapeake Division, Society of Naval Architects and Marine Engineers, Alexandria, Va., December.

Ross, D. (1979), *Energy From the Waves*, Pergamon Press, Oxford, U.K.

Slater, S. H. (1974), "Wave Power," *Nature*, Vol. 249, No. 5459, June.

Salter, S., Jeffrey, D., and Taylor, J. (1976), "The Architecture of Nodding Duck Wave Power Generators," *The Naval Architect*, January, pp. 21–24.

Serman, D. (1978), "Theory of Salter's Wave Energy Device in a Random Sea," M.S. thesis, Department of Civil Engineering, Massachusetts Institute of Technology, Cambridge, Mass.

Skongaard, O., Jonsson, I., and Brink-Kjer, O. (1975), "Scattering of Water Waves at an Idealized Island," Technical University of Denmark, Progress Report 36, August, pp. 33–36.

Streeter, V. L. (1971), *Fluid Mechanics*, 5th ed., McGraw-Hill, New York.

Wendel, K. (1956), "Hydrodynamic Masses and Hydrodynamic Moments of Inertia," David Taylor Model Basin, *Translation* No. 260, July.

Wiegel, R. L. (1964), *Oceanographical Engineering*, Prentice-Hall, Englewood Cliffs, N.J.

Wirt, L. S. (1976), "Harvesting Ocean Wave Energy with the Lockheed DAM–ATOLL," Lockheed California Company, Report LR27803.

Wirt, L., and Higgins, T. (1979), "DAM–ATOLL Ocean Wave Energy Extraction," Marine Technology Society Meeting, New Orleans, La., October 11–12.

5

Energy Conversion, Transmission, and Storage

In Sections 2.1 and 2.2 we assume waves to have single frequencies and unvarying behaviors. These assumptions are also used in the analyses of the various wave energy conversion devices discussed in Chapter 4. In Section 2.3, however, waves are treated as they occur in nature, that is, randomly as in the wind-generated sea and irregularly as in the case of the swell. Because of this time-varying nature of the sea, the direct conversion of wave energy into ac electricity is most difficult. This conversion is necessary if the converted wave energy is to be supplied to the national grid.

The energy requirements of a country can be partially satisfied by ocean waves in ways other than by direct electrical transmission. These methods include conversion to *dc electricity* and the subsequent storage in batteries and the creation of *energy-intensive products*, such as aluminum, which can be transferred by ship. The reader must realize that wave energy is a distributed resource and, as such, must be converted in many regions. Thus the energy can be used directly by a single household or an isolated seacost community. Direct current electricity is acceptable in these situations. On the other hand, wave energy conversion in locations far from land where the resource is relatively constant in time, such as the trade wind belts, is most suitable for the production of energy-intensive products.

In this chapter, several electromechanical energy conversion systems are discussed. The various methods of transmission and storage are also described.

5.1 Basic Electromechanical Energy Conversion Techniques

The eight generic wave energy conversion techniques described in Sections 4.1 and 4.2 (wave focusing is not a conversion technique per se) have one or more possible electromechanical energy conversion techniques associated with them. In this section the two most popular electromechanical systems are described: the *mechanical-drive* generator and the *fluid-drive* generator.

A Mechanically Excited Generators

Consider the wave energy conversion systems sketched in Figure 5.1. In Figure 5.1*a* a heaving float is attached to a line that passes over a *pulley* connected to a rotating electrical generator and then to a weight. The weight ensures that the line remains in tension. As the float rises the pulley turns in a clockwise direction while the pulley direction is reversed as the float falls. A *ratchet* device can be used to insure that the generator rotates in only one direction. A variation of this basic system is to replace the weight by a *spring-loaded* pulley system at the generator, as in Figure 5.1*b*, where a purely pitching body is sketched using a pully generator device similar to that in Figure 5.1*a*.

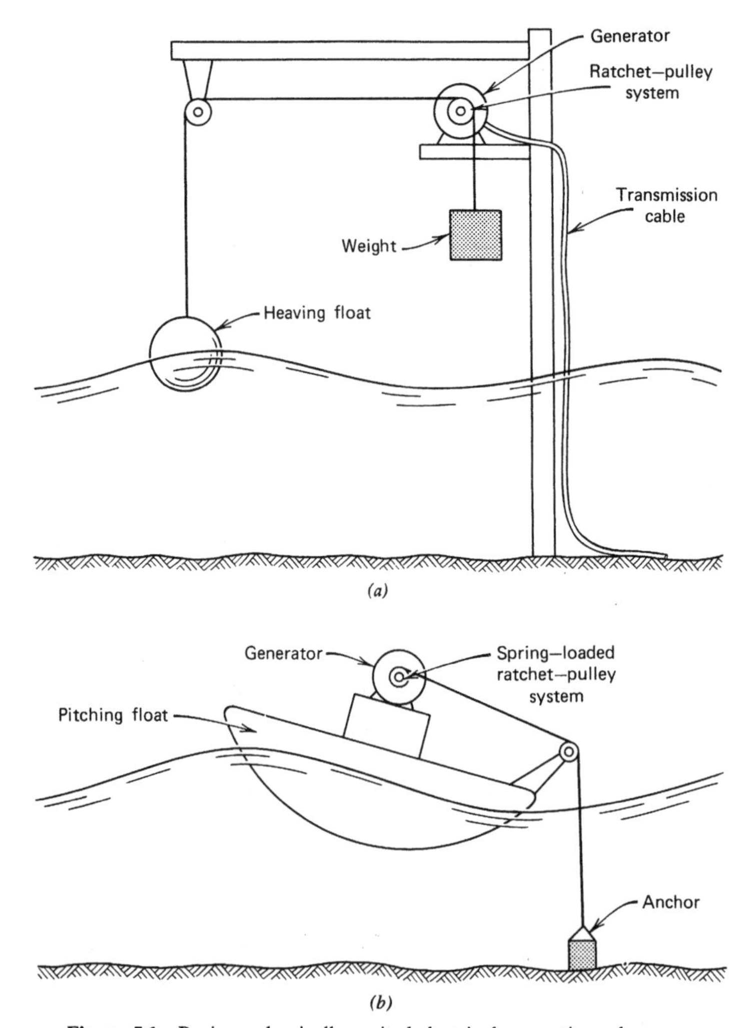

Figure 5.1 Basic mechanically excited electrical generation schemes.

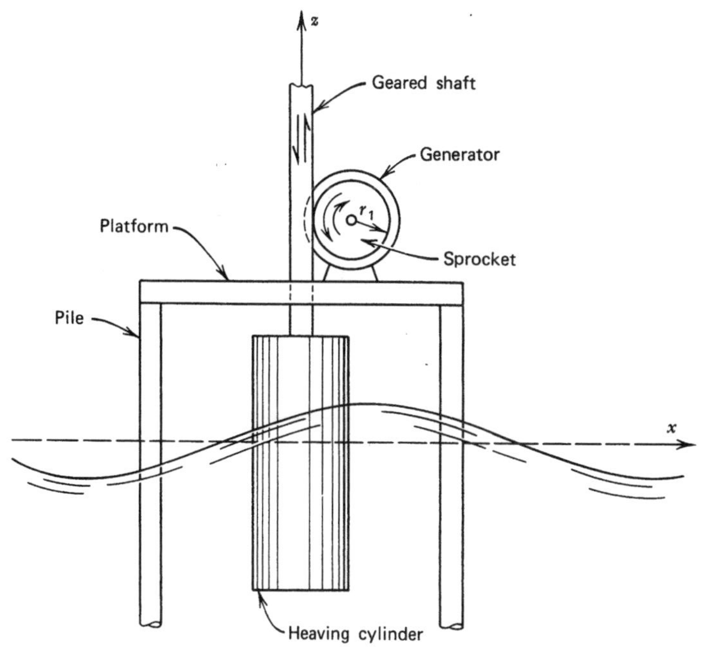

Figure 5.2 Heaving excited gear–sprocket generator system.

Another mechanical-drive system that is often suggested for heaving wave energy converters is the *gear* system sketched in Figure 5.2. In Figure 5.2*a* a heaving circular cylinder is shown attached to a vertical shaft having gears on its right side. These gears mate with a sprocket that, in turn, is attached to the generator shaft. The motions of the cylinder are described by equation (4.12) where, in the case of *resonance*, the phase angles γ and σ_z have values of 0 and 90°, respectively. We assume this condition in the analysis that follows; thus at *resonance*

$$z = Z_0 \sin(\omega t) \tag{5.1}$$

where Z_0 is the heaving amplitude and ω is the circular wave frequency. The vertical velocity of the cylinder and geared shaft is

$$\begin{aligned} v_0 &= \frac{dz}{dt} \\ &= \omega Z_0 \cos(\omega t) \end{aligned} \tag{5.2}$$

Assuming no slippage between the gear and sprocket, the velocity of equation (5.2) is the same as the linear velocity of the perimeter of the sprocket; thus

$$\begin{aligned} v_1 &= r_1 \omega \\ &= v_0 \\ &= \omega Z_O \cos(\omega t) \end{aligned} \tag{5.3}$$

so that

$$\omega_1 = \frac{\omega Z_0}{r_1} \cos(\omega t) \tag{5.4}$$

where r_1 is the radius of the sprocket and ω_1 is the angular speed or, more germanely, the *circular frequency of rotation*. This rotational frequency ω_1 is most important since the power output of a rotational dc generator is a function of this frequency. The result of equation (5.4) illustrates the fact that the generator frequency is time varying, a rather disturbing fact to be accounted for in the electrical design. Our interest is not as much in the instantaneous value of ω_1 as in the *time-averaged value*,

$$\hat{\omega}_1 = 2\pi \hat{f}_1$$

$$= \frac{2\omega Z_0}{\pi r_1} \tag{5.5}$$

since $\hat{f}_1$ is normally the value specified in the electrical design.

Example 5.1

Consider the heaving circular cylinder in Figure 5.2 to be in resonance with a 7-sec wave and having a motion amplitude of 3 ft (0.914 m). The sprocket radius r_1 is 0.5 ft (0.152 m). Thus the average generator frequency of equation (5.5) is

$$\hat{\omega}_1 = \frac{2}{\pi}\left(\frac{2\pi}{T}\right)\frac{Z_0}{r_1}$$

$$= 3.43 \text{ rad/sec}$$

or

$$\hat{f}_1 = \frac{\hat{\omega}_1}{2\pi}$$

$$= 0.546 \text{ Hz}$$

The *desirable frequency range for a dc generator is between 10 and 60* Hz; therefore, the sprocket radius required to achieve this range must be between 0.0273 ft (9.32×10^{-3} m) and 4.55×10^{-3} ft (1.24×10^{-4} m), respectively. These radius values are so small that they are *impractical*.

The results in Example 5.1 show that the gear system illustrated in Figure 5.2 is impractical in wave energy conversion. A more feasible configuration is shown in Figure 5.3, where the single sprocket is replaced by a *compound*

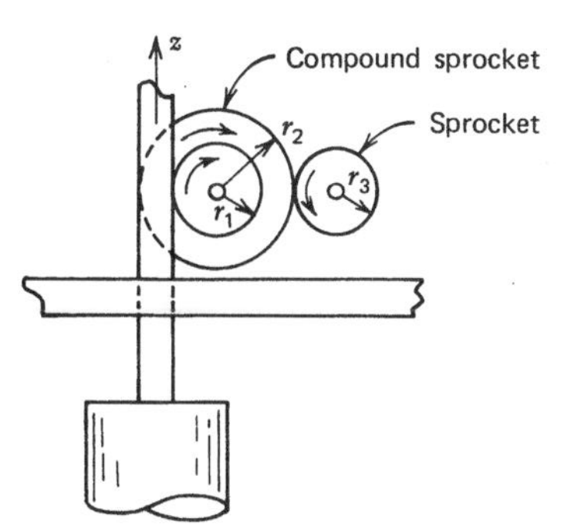

Figure 5.3 A compound gear–sprocket system.

sprocket. In this case wheels of radii r_1 and r_2 are rigidly connected so that the rotational speeds are equal. Thus equations (5.1) through (5.5) still apply, and the rotational speed of the r_2 wheel is also described by equations (5.4) and (5.5). The averaged linear velocities of the r_1 and r_2 wheels are then related by the expression

$$\hat{\omega}_1 = \frac{\hat{v}_1}{r_1} = \frac{\hat{v}_2}{r_2} \tag{5.6}$$

Furthermore, the relationship between the motions of the r_2 and r_3 wheels is

$$\begin{aligned} \hat{v}_2 &= r_2\hat{\omega}_1 \\ &= \hat{v}_3 \\ &= r_3\hat{\omega}_3 \end{aligned} \tag{5.7}$$

where $\hat{\omega}_3$ is, now, the average generator angular frequency. Therefore, using the results of equations (5.5) and (5.7), we obtain

$$\begin{aligned} \hat{\omega}_3 &= \frac{r_2}{r_3}\hat{\omega}_1 \\ &= \frac{2}{\pi}\left(\frac{r_2}{r_1 r_3}\right)\omega Z_0 \end{aligned} \tag{5.8}$$

Example 5.2

Again, consider the wave conditions in Example 5.1 to exist, namely, $T = 7$ sec and $Z_0 = 3$ ft (0.914 m). The heaving cylinder in this case, however, generates electricity using the gear system sketched in Figure 5.3. We require an average generator frequency f_3 of 25 Hz; therefore, using the expression of equation (5.8), we obtain the following expression for the radius of the sprocket directly attached to the generator:

$$r_3 = \frac{2}{\pi}\frac{r_2}{r_1}\frac{\hat{\omega}}{\hat{\omega}_3}Z_0$$

$$= \frac{2}{\pi}\frac{r_2}{r_1}\frac{2\pi}{T}\frac{Z_0}{2\pi f_3}$$

$$= \frac{2}{\pi}\frac{r_2}{r_1}\frac{3}{7(25)}$$

$$= 0.0109\frac{r_2}{r_1}\ \text{(ft)}$$

or

$$r_3 = 3.32 \times 10^{-3}\frac{r_2}{r_1}\ \text{(m)}$$

If $r_2/r_1 = 5$, then r_3 is 0.0546 ft (0.0166 m) or 0.655 in., which is a practical value.

The instantaneous *generator angular frequency* in this case is

$$\omega_3 = \left(\frac{r_2}{r_1 r_3}\right)\omega Z_0 \cos(\omega t) = \omega_{3_0}\cos(\omega t) \tag{5.9}$$

Thus the *maximum generator angular frequency* is simply

$$\omega_{3_0} = \left(\frac{r_2}{r_1 r_3}\right)\omega Z_0$$

$$= 91.7(3)\omega = 247\ \text{rad/sec} \tag{5.10}$$

or

$$f_{3_0} = 39.3\ \text{Hz}$$

which, again, is a practical value.

The *rotating generators* used with the system sketched in Figures 5.1 through 5.3 could be either ac or dc. Since the generator frequency ω_3 in equation (5.9) is sinusoidal in time it is best to use a dc generator with geared systems similar to those in Figures 5.2 and 5.3. The *instantaneous rectified voltage* of such a generator can be represented by

$$e = e_0|\sin(\omega_3 t)|$$

$$= e_0\left|\sin\left[\omega_{3_0}\cos(\omega t)\right]\right| \tag{5.11}$$

where the expression for ω_3 in equation (5.9) is used. The *instantaneous current* i has a form similar to the voltage e, but with a phase difference; hence

$$\begin{aligned} i &= i_0|\sin(\omega_3 t + \Phi)| \\ &= i_0\left|\sin\left[\omega_{3_0}\cos(\omega t)t + \Phi\right]\right| \end{aligned} \tag{5.12}$$

where Φ is the *phase angle*. The *instantaneous power* output of the generator is then

$$P_0 = ei \tag{5.13}$$

The reader can see the complexity introduced by the time-dependent generator frequency ω_3 in the expression for the instantaneous voltage e and current i of equations (5.11) and (5.12), respectively. The problem of wave energy conversion into electrical energy by rotating generators has been discussed by several authors at the Symposium on Wave Energy Utilization held at Chalmers University in Gothenburg, Sweden (Bishop and Rees, 1979; Binns 1979; Thorborg, 1979; Wilson, 1979). It is beyond the scope of this book to cover the electrical engineering aspects of wave energy conversion and utilization in more detail; therefore, the reader is encouraged to consult the aforementioned references.

B Pneumatically and Hydraulically Excited Generators

The wave energy conversion technique using *cavity resonance*, discussed in Section 4.1, B, requires either air or water turbines in the production of electricity. A number of turbine configurations have been suggested and tried, primarily in the pneumatic (air) category. Some of these are described in the publications of Masuda and Miyasaki (1978), Whittaker and Wells (1978), and McCormick (1978). These references describe three radically different air turbines.

The turbine system of Masuda and Miyasaki (1978), sketched in Figure 5.4, is the forerunner of the three turbines. Its operation requires two resonant chambers which, respectively and alternately, create high pressure upstream of the turbine or a partial vacuum downstream. Intake and exhaust flaps are required to maintain unidirectional flow past the turbine. This turbine configuration, theoretically analyzed by Hiramoto (1978), is that used by the Japanese in the "Kaimei" experiment mentioned in Section 4.1, B and sketched in Figure 4.14.

Whittaker and Wells (1978) describe a self-rectifying turbine commonly referred to as the *Wells turbine*. This turbine consists of blades with nonlifting (symmetric) profiles attached to a rotor, as sketched in Figure 5.5. Since the blades are symmetric in profile and have no natural angle of attack, a *prespin* is required to generate both lift and thrust, as shown in Figure 5.5. Obviously, the operation of the turbine is independent of the axial flow direction and,

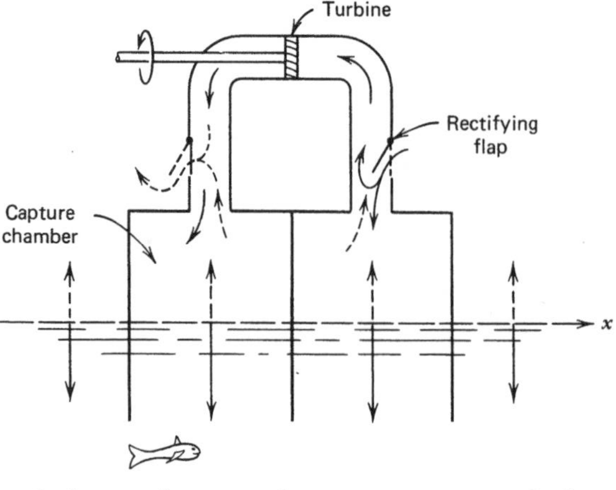

Figure 5.4 Schematic diagram of a pneumatic wave energy convertor having unidirectional flow and rectifying flaps.

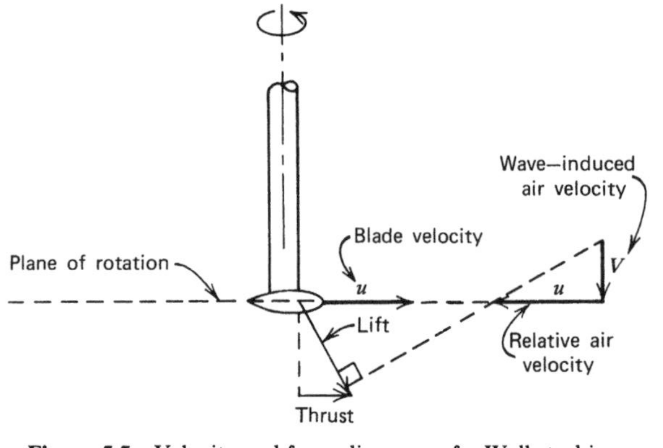

Figure 5.5 Velocity and force diagrams of a Wells turbine.

therefore, has the *self-rectifying* property, requiring none of the flaps shown in Figure 5.4. Only one resonant chamber is, therefore, required.

The air turbine described by McCormick (1978), shown in Figure 5.6, also has the self-rectifying property. Referring to Figure 5.6, the air excited in a single resonant chamber is guided past the upstream rotor, which becomes the primary power rotor. This rotor, in turn, acts as a "moving stator" for the downstream *counterrotating* secondary power rotor. When the airflow direction is reversed, the rotors simply exchange roles. The counterrotating system was conceived by McCormick (1978) for use in the "Kaimei" experiment. The preliminary design was performed by Professor R. Latham of the U.S. Naval Academy, and the final design was performed by J. Schwartz of the David W. Taylor Naval Ship Research and Development Center and Professor W. Lee of the U.S. Naval Academy.

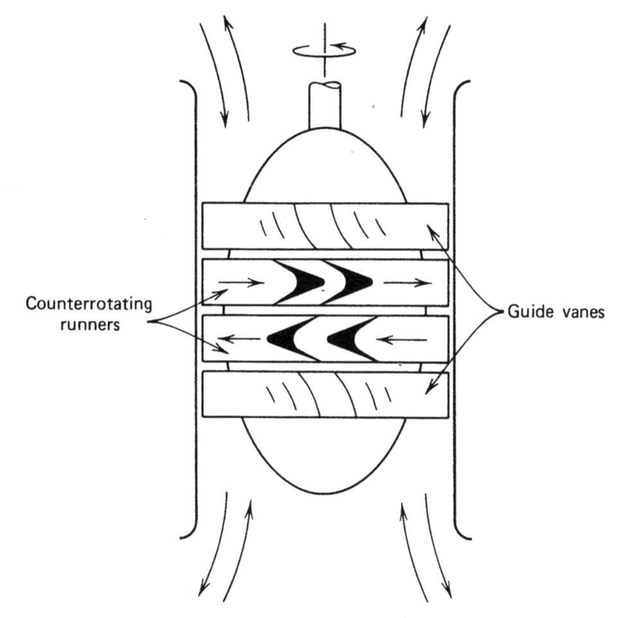

Figure 5.6 The McCormick counterrotating air turbine.

Other self-rectifying turbine systems have been patented by I. A. Babinsten in 1975 (U.S. Patent Number 3,922,729) and by G. D. Filipenco in 1975 (U.S. Patent Number 3,912,932). The Babintsen concept is illustrated in Figure 5.7, and the Filipenco scheme is illustrated in Figure 5.8.

The analysis of the pneumatic turbines is complicated by the fact that the airflow is *unsteady*. In a monochromatic wave the expression for the air power available to the turbine is rather complex since both nonlinear and inertial terms are included. McCormick (1974) presents such a power expression for a pneumatic wave energy conversion device similar to that sketched in Figure 4.9. The results of that expression are shown in Figure 4.10. More recently,

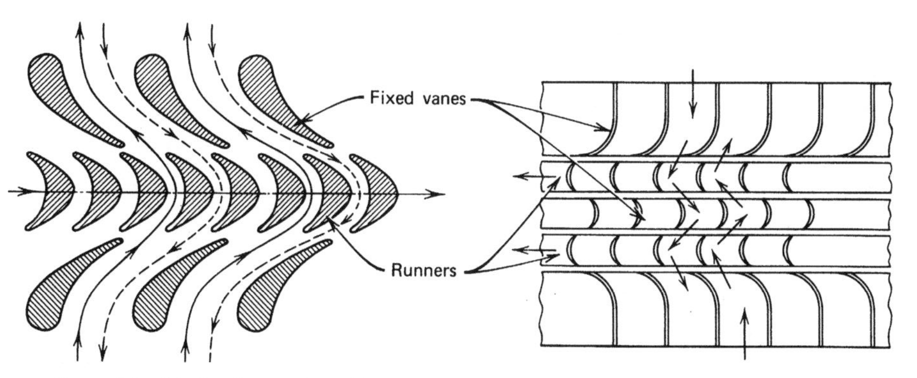

Figure 5.7 The Babinsten air turbine.

Figure 5.8 The Filipenco air turbine.

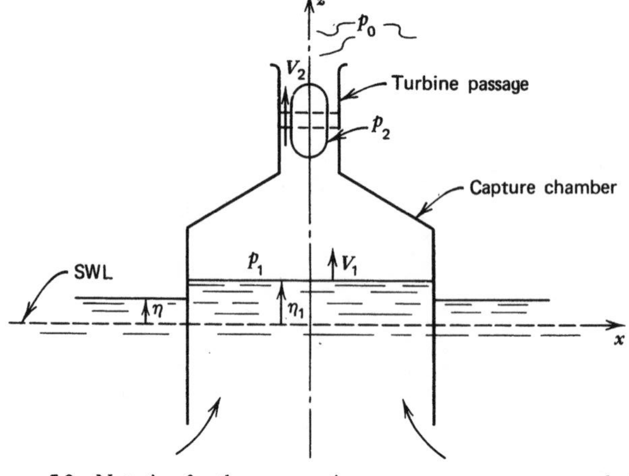

Figure 5.9 Notation for the pneumatic wave energy convertor analysis.

Hiramoto (1978) presented both wave and turbine power results based on a quasi-steady theory that included the effects of air compressibility.

The general method of analysis is outlined here while using the results of the theoretical turbine analysis of Hiramoto (1978). Referring to Figure 5.9 for notation, the internal free-surface displacement from its equilibrium position is

$$\eta_1 = \frac{\overline{H}_1}{2} \cos(\omega t) \tag{5.14}$$

where $\overline{H}_1$ is the spatially averaged internal wave height. The velocity of the air adjacent to the internal free-surface is then

$$\begin{aligned} V_1 &= \frac{d\eta_1}{dt} \\ &= -\frac{\omega \overline{H}_1}{2} \sin(\omega t) \end{aligned} \tag{5.15}$$

and, assuming *incompressible flow*, the *axial velocity* in the *turbine passage* is

$$\begin{aligned} V_2 &= \frac{A_1}{A_2} V_1 \\ &= -\frac{A_1}{A_2} \frac{\omega \overline{H}_1}{2} \sin(\omega t) \end{aligned} \tag{5.16}$$

where A_1 and A_2 are the flow areas of the capture chamber (cavity) and the turbine passage, respectively.

The *power* P available to the turbine depends on the pressure gradient and *volume rate of airflow* Q across the turbine:

$$P = (p_2 - p_0)\, Q \tag{5.17}$$

where, from the equation of *continuity*,

$$Q = V_1 A_1 = V_2 A_2 \tag{5.18}$$

The exhaust pressure p_0 is assumed to be ambient and, for simplicity, constant in the neighborhood of the turbine. Finally, the upstream pressure p_2 is related to the pressure within the air chamber by the *energy equation* due to Bernoulli,

$$p_2 = p_1 + \frac{1}{2}\rho(V_1^2 - V_2^2) + \rho \frac{\partial}{\partial t}(\varphi_1 - \varphi_2) \tag{5.19}$$

where the *velocity potentials* φ_1 and φ_2 are approximated by

$$\varphi_1 \simeq V_1 \eta_1 = -\frac{\omega H_1^2}{4} \sin(\omega t)\cos(\omega t) \tag{5.20}$$

and

$$\varphi_2 \simeq \left(\frac{A_1}{A_2}\right)\varphi_1 \tag{5.21}$$

respectively. Finally, the pressure difference in the power expressed of equation (5.17) is obtained from the *linear momentum equation*:

$$p_2 - p_0 \simeq \rho\left(\frac{A_1}{A_2}\right)\frac{\partial \varphi_1}{\partial t} + \rho \frac{Q}{A_2}(V_2 - V_1) \tag{5.22}$$

Since the motions of the internal free surface, obtained from equation (5.14), are assumed to be known, the pressures p_1 and p_2 can now be determined.

The results of the quasi-steady analysis of Hiramoto (1978) are now presented. The term "quasi-steady" is used here since the time variations of the velocity potentials φ_1 and φ_2 are neglected in the analysis. McCormick (1974) did include these unsteady terms in predicting power available to the turbine. The relatively good agreement between Hiramoto's analytical results and experimental results justifies the neglect of the unsteady terms.

Hiramoto (1978) derives the expression for the power of a pneumatic wave energy conversion turbine with a blade profile similar to that in Figure 5.10. The resulting *turbine power* expression is

$$P_T = \rho Q u_T \left[w_2 \cos(\psi_2) + \sqrt{\frac{w_2^2 + u_T^2 - 2 w_2 u_T \cos(\psi_2)}{1 + Y_T}}\, \sin(\alpha_2) - u_T \right] \tag{5.23}$$

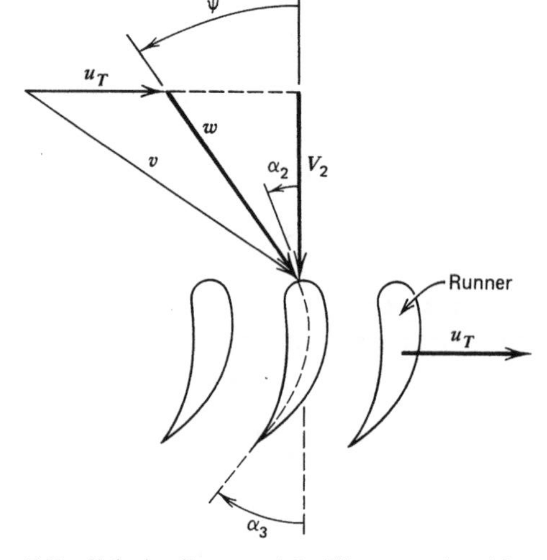

Figure 5.10 Velocity diagram of the Hiramoto air turbine analysis.

where ψ_2 is the turning angle of the upstream guide vanes, α_2 is the blade angle, w_2 is the absolute inlet velocity, and Y_T is the *blade coefficient* defined by

$$Y_T = \left(\frac{v_2}{v_3}\right)^2 - 1 \tag{5.24}$$

where v_2 and v_3 are the inlet and exhaust relative air velocities, respectively. The linear runner velocity u_T is obtained by using the angular rotor speed ω_T and the midspan rotor radius; thus

$$\bar{r} = \frac{r_i + r_0}{2} \tag{5.25}$$

where r_i is the *hub radius* and r_0 is the *blade tip radius*. The linear runner velocity is then

$$u_T = \bar{r}\hat{\omega}_T \tag{5.26}$$

where $\hat{\omega}_T$ is the time-averaged value of the angular rotor velocity. Whereas ω_T is a function of time, the variation from the average value is minor since the *inertia* of the rotating system gives the system an *energy storage* capability. Since the value of $\bar{\omega}_T$ is a *design value*, u_T is known and the relative velocities of equation (5.24) are

$$v_2 = \sqrt{V_2^2 + (u_T + V_2 \tan(\psi_2))^2} \tag{5.27}$$

$$v_3 = \frac{V_2}{\cos(\alpha_3)} \tag{5.28}$$

from the velocity vector diagram in Figure 5.10. Finally, the *inlet absolute velocity* is

$$w_2 = \frac{V_2}{\cos(\alpha_2)} \tag{5.29}$$

All quantities in the power expression of equation (5.23) are known. Hiramoto (1978) applied the analysis to the Japanese turbine in the Kaimei experiment, which had the following properties and design characteristics:

$$A_1/A_2 = 118$$

$$A_2 = 4.20 \text{ ft}^2 \ (0.39 \text{ m}^2)$$

$$\overline{H} = 9.84 \text{ ft} \ (3.00 \text{ m})$$

$$r_0 = 2.30 \text{ ft} \ (0.700 \text{ m})$$

$$T = 6 \rightarrow 7 \text{ sec}$$

$$P_T = 125 \text{ kW}$$

Comparing the pneumatically excited generator to the mechanically excited generator of Section 5.1, A, the reader can see that the pneumatic device is superior since the operation is *continuous*. Furthermore, the *high rotational speeds* desirable for electrical generation are obtainable in a pneumatic system.

5.2 Advanced Electromechanical Energy Conversion Techniques

The energy conversion techniques described in Section 5.1 have been suggested by approximately 95% of the inventors with whom the author is acquainted. Many of the system components required by these techniques are "off-the-shelf" items, and, therefore, the associated costs are relatively low. In this section three nonstandard energy conversion ideas are discussed. Since their development is somewhat in the future, and because the components can be totally isolated from the weather, these techniques are referred to as "advanced."

A Linear Inductance

A *linear inductance generator* simply consists of either a permanent magnet moving through a coil or vice versa. The concept is not new; however, its application to wave energy conversion is rather recent. One of the best descriptions of a linear inductance wave energy conversion device is in the paper by Omholt (1978). Referring to Figure 5.11, a conducting wire loop is situated between two permanent magnets that are part of a mass suspended

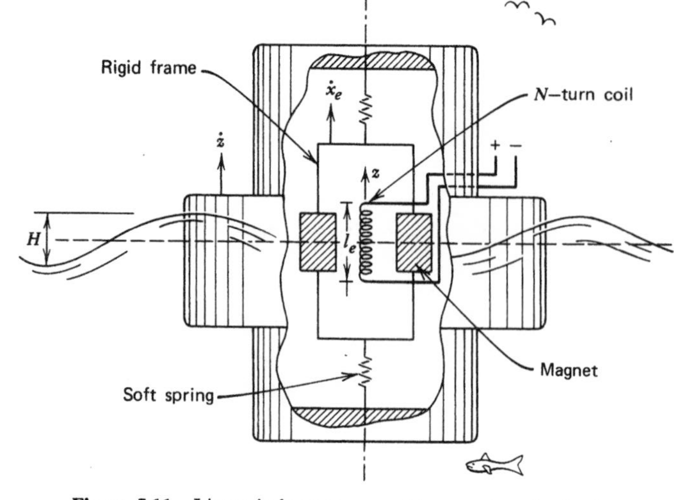

Figure 5.11 Linear inductance generator on a heaving body.

by a spring system. The wire loop is fixed the respect to a heaving body excited by the waves. The springs are rather "soft" and, therefore, isolate the motions of the magnets from those of the wire loop–heaving body system. Electrical current is produced in the wire loop as a result of the relative motion between the wire loop and the magnet.

From the analysis of Omholt (1978), the *electrical power* produced by the system sketched in Figure 5.11 is obtained from

$$P_e = \frac{N_e^2 B_e^2 l_e^2 \dot{\xi}^2}{R_e} \tag{5.30}$$

where N_e is the number of turns of the wire loop, B_e is the magnetic induction, l_e is the length of the wire within the magnetic field, and R_e is the load resistance in the wire loop. The generation of the electrical power results in a motion *resisting force*, the magnitude of which is obtained from

$$F_e = \frac{P_e}{\dot{\xi}} \tag{5.31}$$

where the relative displacement ξ and the velocity $\dot{\xi}$ are, respectively,

$$\xi = z - x_e \tag{5.32}$$

and

$$\dot{\xi} = \dot{z} - \dot{x}_e \tag{5.33}$$

Referring to Figure 5.11, the *heaving body* (and *wire loop*) *displacement* from equation (4.12) is

$$z = Z_0 \cos(\omega t + \gamma - \sigma_z) \tag{4.12}$$

and that of the magnets is

$$x_e = x_0 \cos(\omega_e t + \varphi_e) \tag{5.34}$$

We assume that the heaving body or float is both in *resonance* with the wave, so that $\sigma_z = 90°$, and is symmetric about a vertical axis, so that $\gamma = 0$ in equation (4.12). Furthermore, we design the systems such that the circular frequency of the magnet motion is equal to that of the wave:

$$\begin{aligned} \omega_e &= \sqrt{\frac{k_e}{m_e}} \\ &= \omega \end{aligned} \tag{5.35}$$

where k_e is the *spring constant* and m_e is the mass of the magnet system. Omholt (1978) also specifies that the buoy motions should lead the motions of magnets by 90°; thus the phase angle of equation (5.34) is

$$\varphi_e = -180° \tag{5.36}$$

This value of the phase angle indicates that the motions of the magnets are 180° out of phase with the wave. We can now combine the results of equations (5.33) through (5.36) and obtain the relative velocity under the condition of resonance; that is,

$$\dot{\xi} = \omega\left[Z_0 \cos(\omega t) - X_0 \sin(\omega t) \right] \tag{5.37}$$

The *time-averaged power* from the linear inductance device over one cycle is obtained from the combination of equations (5.30) and (5.37):

$$\begin{aligned} \hat{P}_e &= \frac{1}{T}\int_0^T P_e \, dt \\ &= \frac{N_e^2 B_e^2 l_e^2}{R_e}\left(\frac{4}{T}\right)\int_0^{T/4} \dot{\xi}^2 \, dt \\ &= \frac{N_e^2 B_e^2 l_e^2 \omega^2}{R_e}\left(\frac{4}{T}\right)\int_0^{T/4} \left[Z_0 \cos(\omega t) - X_0 \sin(\omega t) \right]^2 dt \\ &= \frac{N_e^2 B_e^2 l_e^2 \omega^2}{2R_e}\left[Z_0^2 - \frac{4 Z_0 X_0}{\pi} + X_0^2 \right] \end{aligned} \tag{5.38}$$

Example 5.3

The heaving body in Example 4.3 is in resonance with a wave having a period of 2.22 sec ($\omega = 2.83$ rad/sec). The resonant heaving amplitude is $Z_0 = 1$ ft (0.305 m).

On the heaving body is a linear inductance generator that has a 100-turn wire loop ($N_e = 100$) that is 2 ft (0.610 m) in length ($l_e = 2$) and a magnetic induction of 1.00 Wb/ft^2 (10.8 Wb/m^2), that is, $B_e = 1$. For simplicity, it is best to work in *international units* in this case since the power expressions of equations (5.30) and (5.38) are then in watts. Assume that the amplitude of magnet motion X_0 is 1 ft (0.305 m). Also, the load resistance in the circuit is 5 Ω ($R_e = 5$). Using these property values, the average power per cycle, obtained from equation (5.38), is

$$\hat{P}_e = \frac{N_e^2 B_e^2 l_e^2 \omega^2}{2R_e}\left[Z_0^2 - \frac{4Z_0 X_0}{\pi} + X_0^2\right]$$

$$= \frac{(100)^2(10.8)^2(0.610)^2(2.83)^2}{2(5)}[0.0676]$$

$$= 23{,}500 \text{ W } (23.5 \text{ kW})$$

The *average mechanical power* in the heaving motion obtained in Example 4.3 is

$$\hat{P}_z = 31{,}700 \text{ W } (31.7 \text{ kW})$$

Thus, for the conditions described in this example, the linear induction system is *74.1% efficient.*

Some *caution* must be taken in choosing the parametric design values in the linear induction energy conversion system. For example, to ensure resonance with a swell, the condition given in equation (5.35) must be satisfied. This means that the magnetic mass and spring must be properly matched. The mass of a magnetic m_e that has the inductance B_e needed for significant power is rather large. Thus the choice of a spring of constant k_e may be difficult. This problem can be circumvented by varying the other design parameters in equations (5.30) and (5.38). Specifically, the power output can be increased by increasing the number of wire turns and the length of the wire loop while decreasing the load resistance.

Work on the linear inductance wave energy conversion system is also in progress at the Langley Research center in Hampton Field, California by D. C. Grana and R. T. Wilem. This work is briefly described in an article in the June 1978 issue of Mechanical Engineering (Anonymous, 1978).

B Piezoelectricity

Piezoelectric crystals have many applications where the conversion of the energy of pressure fluctuations to electrical energy is required. One of the most common piezoelectric devices is the hydrophone, where the small hydroacoustic pressures are the energy source. Burfoot and Taylor (1979) present an excellent discussion of the theory and applications of piezoelectricity. The second author of that 1979 publication also recognized the possible application of piezoelectricity to water wave energy conversion. Under a U.S. Department of Energy contract he performed an experimental study in a wave tank, the results of which are presented in the publication by Taylor (1979). The materials presented in this section are based on the Taylor study.

Piezoelectric materials are of two types: *transverse* and *compressional.* Materials that produce a voltage difference in a direction perpendicular to the direction of the pressure force are called *transverse materials*. When the voltage difference is in the direction of the pressure force, the material is called a *compressional material.* For wave energy conversion the compressional piezoelectric materials are preferred since "stacking" then is possible, as illustrated in Figure 5.12. In Figure 5.12 a thin piezoelectric film is positioned on a rigid supporting structure. These structures can be stacked as sketched in Figure 5.12*b*, enabling many films to be positioned in a relatively small surface area. If the *design wave* is of length λ and height H we require that *stack length* be

$$L_{pc} \ll \lambda \tag{5.39}$$

and the *stack separation* be

$$\Delta_{pc} \simeq H \tag{5.40}$$

so that the energy conversion structure does not significantly reflect the wave.

Following the analysis of Taylor (1979), the *mechanical energy* of the wave-induced pressure on the piezoelectric film is

$$E_m = \frac{p^2 l_{pc} b_{pc} \delta_{pc}}{Y_{pc}} \tag{5.41}$$

where, referring to Figure 5.12*a*, p is the pressure and Y_{pc} is the modulus of elasticity. The energy conversion ability of the material is characterized by both the respective *piezoelectric stress* and *strain constants*

$$g_{pc} = \frac{\partial e}{\partial p} \tag{5.42}$$

in volt-feet per pound or volt-meters per Newton where e is the instantaneous

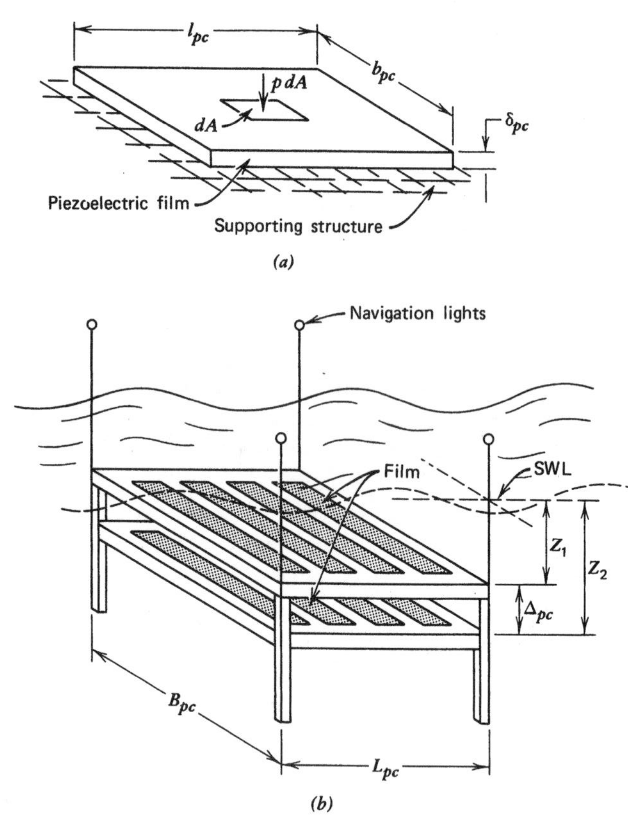

Figure 5.12 Piezoelectric film orientation for wave energy conversion: (*a*) notation for piezoelectric wave energy conversion analysis; (*b*) a two-layer piezoelectric conversion stack.

voltage and

$$d_{pc} = \epsilon_{p0}\, g_{pc} \tag{5.43}$$

in feet per volt or meters per volt.

***Note*:** Most of the recent data available on piezoelectric materials are given in terms of international (SI) units; thus we use these units exclusively in this section. In equation (5.43) ϵ_{pc} is called the *dielectric permittivity* of the material, the units of which are pounds per square or newtons per square volt. The corresponding electrical energy produced by the piezoelectric material is

$$E_{pc} = E_m(Y_{pc} g_{pc} d_{pc})$$

$$= p^2 l_{pc} b_{pc} \delta_{pc} g_{pc} d_{pc} \tag{5.44}$$

Examples of transverse and compressional materials, from Taylor (1979), are presented in Table 5.1.

Table 5.1 Piezoelectric Material Properties

Material	Type	Y_{pc}(N/m^2)	g_{pc}(V-m/N)	d_{pc}(m/V)
Kureha Piezofilm	Transverse	2.9×10^9	0.200	2.3×10^{-11}
Honeywell Composite	Compressional	3.8×10^7	0.560	34×10^{-11}

Example 5.4

A two-layered piezoelectric stack similar to that in Figure 5.12b is positioned 6 ft (1.83 m) below the SWL. The average swell at this location has a 10-sec period and a 3-ft (0.914-m) height. The depth of water h is 30 ft (9.14 m); thus the depth : deep water wavelength ratio is

$$\frac{h}{\lambda_0} = 0.0585$$

Using this value in Figure 2.2, we see that the condition is approximately that of *shallow water*. We can, therefore, use the expression in equation (2.8) to determine the wavelength at the energy conversion site; thus

$$\lambda = \sqrt{gh}\,T$$

$$= \sqrt{32.2(30)}\,(10)$$

$$= 311 \text{ ft } (94.7 \text{ m})$$

To satisfy the design conditions of equations (5.39) and (5.40), we specify

$$L_{pc} = \frac{\lambda}{20} = 15.6 \text{ ft } (4.74 \text{ m})$$

and

$$\Delta_{pc} = 3.00 \text{ ft } (0.914 \text{ m})$$

Also, we specify the platforms to be square (i.e., $B_{pc} = L_{pc}$).

For this project the transverse material Kureha Piezofilm, which has the properties listed in Table 5.1, is used; thus

$$Y_{pc} = 2.9 \times 10^9 \text{ (N/m}^2\text{)}$$

or

$$Y_{pc} = 6.05 \times 10^7 \text{ (lb/ft}^2\text{)}$$

and

$$g_{pc} = 0.200 \text{ (V-m/N)}$$

and

$$g_{pc} = 2.92 \text{ (V-ft/lb)}$$

and

$$d_{pc} = 2.3 \times 10^{-11} \text{ (m/V)}$$

or

$$d_{pc} = 7.54 \times 10^{-11} \text{ (ft/V)}$$

The thickness of the material δ_{pc} is 1.64×10^{-4} ft (5.00×10^{-5} m) or 50 μm. The sheets of materials are positioned such that the sums ($\sum$) of their lengths and widths approximately equal the platform dimensions; thus

$$\sum l_{pc} \simeq L_{pc} = 15.6 \text{ ft (4.74 m)}$$

and

$$\sum b_{pc} \simeq B_{pc} = 15.6 \text{ ft (4.74 m)}$$

The wave-induced pressure in this shallow water situation is obtained from equation (4.48), where

$$\frac{\cosh(kz + kh)}{\cosh(kh)} \simeq 1 \tag{5.45}$$

Thus

$$p \simeq \frac{\rho g H}{2} \cos(kx - \omega t) - \rho g z \tag{5.46}$$

Only the *dynamic component of pressure* [the first term on the right-hand side of equation (5.46)] excites the piezoelectric material; therefore, the exciting pressures on both piezoelectric layers are

$$p_1 = p_2 = \frac{\rho g H}{2} \cos(\omega t)$$

$$= \frac{2.00(32.2)3}{2} \cos\left(\frac{2\pi}{10} t\right)$$

$$= 96.6 \cos(0.628t) \text{ lb/ft}^2$$

or

$$p_1 = p_2 = 4620 \cos(0.628t) \text{ N/m}^2$$

where the position of the stack is assumed to be at $x = 0$. The pressure p_1 acts on the upper layer located at $z_1 = -6$ ft (-1.83 m), whereas p_2 acts on the lower layer at

$$\begin{aligned} z_2 &= z_1 - \Delta_{pc} \\ &= -9 \text{ ft } (-2.74 \text{ m}) \end{aligned}$$

Using these wave-induced pressures, we now determine the magnitude of the energy converted by the piezoelectric stack by using the results of equation (5.44); thus

$$\begin{aligned} E_{pc} &= 2p_1^2\left(\sum l_{pc}\right)\left(\sum b_{pc}\right)\delta_{pc}g_{pc}d_{pc} \\ &= 2\left[96.6\cos(0.628t)\right]^2(15.6)^2 1.64 \times 10^{-4}(2.92)7.54 \times 10^{-11} \\ &= 1.64 \times 10^{-7}\cos(0.628t) \text{ lb-ft} \end{aligned}$$

or

$$E_{pc} = 2.21 \times 10^{-7}\cos(0.628t) \text{ N-m}$$

The *average energy* converted per wave is

$$\begin{aligned} \hat{E}_{pc} &= \frac{4}{T}\int_0^{T/4} E_{pc}\, dt \\ &= \frac{8}{T}\left(\sum l_{pc}\right)\left(\sum b_{pc}\right)\delta_{pc}g_{pc}d_{pc}\int_0^{T/4} p_1^2\, dt \\ &= \frac{8}{10}(15.6)^2 1.64 \times 10^{-4}(2.92)7.54 \times 10^{-11}\int_0^{2.5}(96.6)^2\cos^2(0.628t)\, dt \\ &= 0.20 \times 10^{-8} \text{ lb-ft} \end{aligned}$$

or

$$\hat{E}_{pc} = 1.10 \times 10^{-7} \text{ N-m}$$

The total energy within the shallow water wave is obtained from equation (2.17) using the wavelength expression of equation (2.8); thus

$$\begin{aligned} E &= \frac{\rho g H^2 \lambda B_{pc}}{8} \\ &= \frac{\rho g H^2 \sqrt{gh}\, I B_{pc}}{8} \\ &= \frac{2.00(32.2)3^2\sqrt{32.2(30)}\ 10(15.6)}{8} \\ &= 3.51 \times 10^5 \text{ lb-ft} \end{aligned}$$

or

$$E = 4.74 \times 10^5 \text{ N-m}$$

The *efficiency* of the piezoelectric stack is then

$$\frac{\hat{E}_{pc}}{E} = \frac{8.20 \times 10^{-8}}{3.51 \times 10^5}$$

$$= 3.24 \times 10^{-13}$$

From the results in Example 5.4 the reader can conclude that piezoelectric materials are not yet feasible for wave energy conversion. Even with substantial technological improvements, there does not appear to be much hope for this direct energy conversion method. Taylor (1979) predicts that breakthroughs in piezoelectric material technology will result in a conversion efficiency of 1%, at best.

C Protonic Conduction

In 1978 Professor Robert E. Salomon introduced the wave energy conversion technique called *protonic conduction*. See the publications of Salomon (1978) and Salomon and Harding (1979). The technique utilizes an *electrochemical hydrogen concentration cell* with *protonic conductors*. Referring to the schematic diagram in Figure 5.13, two chambers containing molecular hydrogen are separated by a material, that allows the conduction of protons only when the concentrations of hydrogen in the two cells differ. Molecular hydrogen H_2 consists of two hydrogen atoms, each having a single proton and a single electron. As the protons pass through the conductor, the electrons deposit on the electrodes, resulting in an electrical current in the external circuit. For the scheme shown in Figure 5.13, the chamber is fixed with respect to the SWL. The hydrogen in the lower chamber is at a constant pressure p_3 since its volume is constant; however, the pressure of hydrogen in the upper chamber p_u varies as a result of the wave-induced piston motions. Thus the pressure of the hydrogen in the upper chamber varies with time;

$$p_u = p_u(t)$$

The reader can see that the gas concentration difference across the protonic conductor is a result of the varying volume of the upper hydrogen chamber.

Following the analysis of Salomon (1978), the *voltage* created by the gas concentration difference is given by the *Nernst equation*,

$$e = \frac{RT_a}{2F} \ln \frac{p_u}{p_l} \tag{5.48}$$

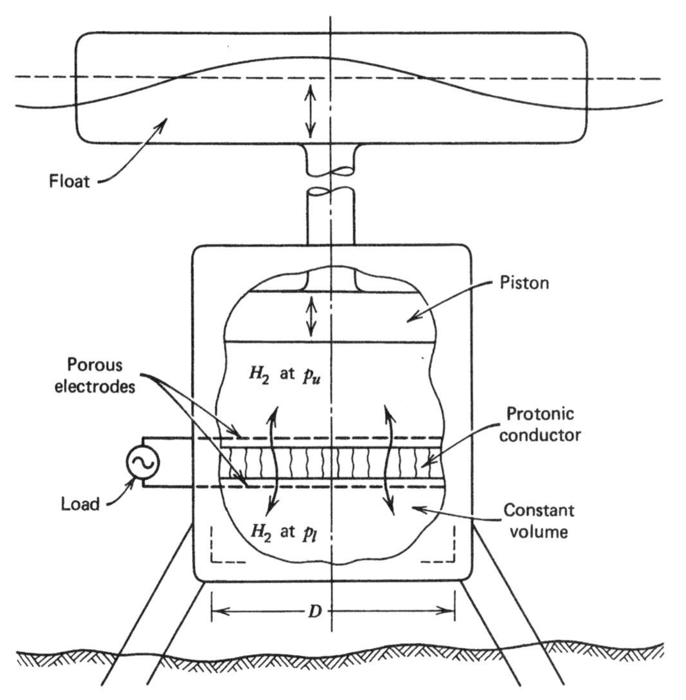

Figure 5.13 Protonic wave energy conversion system with heaving excited piston motions.

where, again, the pressure p_l is constant but p_u varies with time. In equation (5.48), R is the gas constant, T_a is the absolute temperature, and F is the Faraday constant.

There are a number of variations of the scheme shown in Figure 5.13. These variations, thoroughly discussed by Salomon (1978), are shown to improve the energy conversion efficiency of the protonic conduction device. Salomon (1978) also makes several points concerning the efficiency that must be considered. First, there is an *optimum amount of hydrogen* for a given system. An increase in the hydrogen mass beyond this value decreases the efficiency. Furthermore, an adequate seal must be used to prevent hydrogen loss. The second point concerns the *resistance* to conduction of the protons in the electrolyte (conductor). This is an area for future developments; although from the experimental results of Salomon (1978), the material called "Nafion" appears to have promise. One method of overcoming the large resistance of a protonic conductor is by *magnifying the pressure gradient*. This is accomplished by using a scheme similar to that illustrated in Figure 4.16*a*, that is, by having the wave-induced pressure $p(t)$ act over a float waterplane area A_{wp}, which is much larger than that of the conductor $\pi D^2/4$. Since the vertical force due to the wave is the same on the float and piston, the pressure p_u is magnified according to the equation

$$p_u(t) = p(t)\left(\frac{4\Delta_{wp}}{\pi D^2}\right) \tag{5.49}$$

Example 5.5

Salomon (1978) gives an example of a *high-pressure cell* where

$$\frac{\pi D^2}{4} = 9.9 \times 10^{-3} A_{wp}$$

and the mean wave height H is 3.28 ft (1 m). The efficiency of the high-pressure system is

$$\begin{aligned}\frac{\hat{E}_p}{E} &= \left(\frac{\pi D^2}{4A_{wp}}\right)\frac{p_0}{\rho g H}\ln\left[1 + \left(\frac{4A_{wp}}{\pi D^2}\right)\frac{\rho g H}{p_0}\right] \\ &= 0.0992 \ln(1 + 10.1) \\ &= 0.239 \quad \text{or} \quad 23.9\%\end{aligned} \tag{5.50}$$

In equation (5.50) p_0 is the absolute atmospheric pressure; that is, 2117 lb/ft^2 (1.01×10^5 N/m^2). To obtain this efficiency, Salomon (1978) states that a 600-fold reduction in required protonic conductivity is obtained by using the high-pressure technique.

Protonic conduction applied to ocean wave energy conversion is still in its infancy. We look to Professor Robert Salomon and his colleagues at Temple University for further developments in this area.

5.3 Power Scaling

In one of the most important phases in the development of a wave energy conversion system, the *experimental phase*, the engineer studies the performance characterisitcs of a *scaled model* of a system component or of the entire wave energy conversion system. For example, a quarter-scale model of the U.S. turbine designed for the Kaimei wave energy conversion system (Figure 4-14) was extensively tested in the Naval Academy Model Basin (NAMB). The significance of the test results in the *prototype* design must then be determined.

When the engineer refers to the "scale model," he or she is referring to a model that has dimensions equal to the prototype dimensions multiplied by a *scale factor*. To illustrate, the prototype of the U.S. turbine referred to in the last paragraph has an air passage internal diameter of 3.28 ft (1.00 m);

therefore, the quarter-scale model has a diameter of 0.82 ft (0.25 m). Thus if L_p is the prototype length and L_m is the model length, the *scale factor* is

$$n \equiv \frac{L_m}{L_p} < 1 \tag{5.51}$$

The question to be answered is "How do we relate length scaling to power scaling?" A quarter-scale turbine model (length-scaled) does not produce one-quarter of the prototype power. To determine the power scaling, we must use two dimensionless parameters associated with wave mechanics. The first is the *Froude number* defined as $V/(gL)^{1/2}$ where V is a characteristic velocity. This dimensionless number is used extensively in the design of marine vehicles. See, for example, the book by Bhattacharyya (1978). In the design of a model, therefore, the engineer requires that the Froude number for both model and prototype be the same; that is,

$$\frac{V_m}{\sqrt{gL_m}} = \frac{V_p}{\sqrt{gL_p}} \tag{5.52}$$

or, since the gravitational constant g is the same in both cases, the velocity ratio is

$$\frac{V_m}{V_p} = \sqrt{\frac{L_m}{L_p}} = \sqrt{n} \tag{5.53}$$

The results of equation (5.53) show that the *scale factor for velocity* is equal to the square root of the length scale factor n.

The second dimensionless number used in model design is the *Strouhal number* fL/V, where f is the wave frequency. This parameter is used in the analysis of wave-induced vibrations of offshore structures, as described by McCormick (1973) and others. In our model design we require

$$\frac{f_m L_m}{V_m} = \frac{f_p L_p}{V_p} \tag{5.54}$$

or, since $f = 1/T$, where T is the wave period,

$$\frac{L_m}{T_m V_m} = \frac{L_p}{T_p V_p} \tag{5.55}$$

Equation (5.55) can be used to obtain the *time-scaling* relationship,

$$\frac{T_m}{T_p} = \frac{L_m}{L_p}\left(\frac{V_p}{V_m}\right) = n^{1/2} \tag{5.56}$$

The *power expression* for a hydromechanical system can be expressed (for dimensional purposes) as

$$P = C\rho A V^3 \tag{5.57}$$

where A is the flow area and C is a dimensionless constant. Since both C and ρ are assumed to be the same for both the model and the prototype, the *power scaling* is obtained from

$$\begin{aligned}\frac{P_m}{P_p} &= \frac{A_m}{A_p}\left(\frac{V_m}{V_p}\right)^3 \\ &= \left(\frac{L_m}{L_p}\right)^2\left(\frac{V_m}{V_p}\right)^3 \\ &= n^2(n^{3/2}) \\ &= n^{7/2}\end{aligned} \tag{5.58}$$

using the results of equations (5.51) and (5.53) To determine the significance of equation (5.56) and (5.58) in wave energy conversion model analysis, consider Example 5.6.

Example 5.6

Each of the 10 turbines of a Kaimei wave energy conversion system is designed to deliver 125 kW of power in a 6 sec, 3.28-ft (1-m) wave. The inside diameter of the turbine passage of the U.S. turbine is 3.28 ft (1.00 m). Tests conducted in the NAMB involved a quarter-scale model. Thus for the NAMB tests, equations (5.51), (5.56), and (5.58) yield

$$\frac{L_m}{L_p} = n = \frac{1}{4},$$

$$\frac{T_m}{T_p} = n^{1/2} = \frac{1}{2},$$

and

$$\frac{P_m}{P_p} = n^{7/2} = \frac{1}{128},$$

respectively. Thus the wave period used in the model tests is

$$\begin{aligned}T_m &= \frac{T_p}{2} \\ &= \frac{6}{2} \\ &= 3 \text{ sec}\end{aligned}$$

with a corresponding wave frequency of

$$f_m = \frac{1}{T_m}$$

$$= 0.33 \text{ Hz}$$

The scaled wave height is

$$H_m = \frac{H_p}{4}$$

$$= \frac{3.28}{4}$$

$$= 0.82 \text{ ft } (0.25 \text{ m})$$

and the scaled power is

$$P_m = \frac{P_p}{128}$$

$$= \frac{125}{128}$$

$$= 0.976 \text{ kW}$$

The scaling equations can be applied to the actual *wave power* expression to obtain the same results; that is, the scaled wave power is the full-scale wave power divided by 128. Concerning the wave itself, the reader can simply remember that the *wave steepness* must be the same for both the model and prototype; thus

$$\frac{H_m}{\lambda_m} = \frac{H_p}{\lambda_p} \tag{5.59}$$

This expression can be shown to be valid under both deep and shallow water conditions.

5.4 Energy Transmission and Storage

A major problem area in wave energy conversion is that of *energy transfer* from the conversion device to the shore. The first energy transmission technique that comes to mind is electrical transmission by *direct cable*. This technique is certainly the most cost effective over relatively short distances and in water that is not too deep. Unfortunately, if we look to wave energy to fill a significant portion of our energy needs, we must consider waves in the deep open waters since the wave resource is much greater in these waters than in coastal regions. The shoaling processes, including bottom friction, reduce the energy of the deep water wave by up to 90% by the time the wave enters the surf zone. Thus to transport the energy converted from deep water waves, techniques other than electrical cable transmission must be used.

In this section both *electrical cable transmission* and *energy-intensive product transfer* are discussed. Much of the information concerning energy-intensive products has been obtained in connection with the *ocean thermal energy conversion* (OTEC) program of the U.S. Department of Energy. It is interesting to note that the nature of these products also makes them excellent for *energy storage*. For example, in Section 5.4, B the "floating battery" is described. In this case wave energy is converted to, and stored as, electrochemical energy. The barge supporting the floating battery can be towed from the energy conversion point to any market close to a navigable estuary.

A Electrical Cables

The transmission of electrical energy converted from wave energy up to a distance of 18.6 miles (30 km) to an onshore terminal is most easily accomplished by ac cable. Within this distance the depth is normally less than 1500 ft (457 m) and, according the Winer (1975), offers no significant cable laying problems. Since the converted electricity is ac it can be supplied directly to the *power grid*. Concerning the cable, Garrity and Morello (1979) suggest the use of a self-contained oil-filled cable for short-distance transmission of 138 kV ac from a 100-MW ocean thermal power plant. This cable would, then, be suitable for transmission from a 100-MW wave energy conversion power plant.

For transmission *distances greater than 18.6 miles (30 km) but less than 80 miles (129 km) dc cable transmission* is more feasible. As Winer (1975) points out, long-line transmission of dc is more efficient and less costly since only two conductors are required, as opposed to three conductors for ac transmission. Garrity and Morello (1979) recommend that the long-line dc transmission performed by paper-impregnated cables at ± 250 kV.

Both ac and dc transmission by submarine cable have similar design and construction considerations. First, *transmission losses* should be less that 10% of the rated power. These losses are reduced by increasing the conductor size, thus increasing the costs. Second, the cables should be *buried* from 3 to 6 ft (0.914 to 1.83 m) into the sea bed for protection. Naturally, this requirement becomes more expensive as the depth of water increases. Third, as few *cable joints* as possible should be used since these are the weakest points, both mechanically and electrically, in the cable system. Long-length cables up to 6 miles (10 km) are available but costly. Finally, the *riser* from the cable to the wave energy conversion platform must be designed to withstand severe sea conditions. Considerable information is available concerning riser design, particularly in the design of OTEC power plants. For example, see the publications Pieroni et al. (1979), Oliver and Jawish (1979), and Bamford et al. (1978).

Example 5.7

Garrity and Morello (1979) recommend an ac cable transmission from an OTEC power plant to Puerto Rico, where a 100-MW plant could be located 2.3 miles (3.7 km) off shore in 3940 ft (1200 m) of water. Four cables are recommended that are oil-filled and deliver 100 MW at 138 kV. The cables are to be buried over 0.932 mile (1.5 km) of the transmission length. The same transmission scheme could be applied to a wave energy conversion system similar to that sketched in Figure 5.14. A linear array of resonant wave energy conversion devices, such as those described in Chapter 4, are connected to a floating terminal. Since the devices are resonators, they act as antennas and focus the wave energy on themselves as described at the beginning of Section 4.2, D. The electrical energy from the wave energy conversion devices to the floating

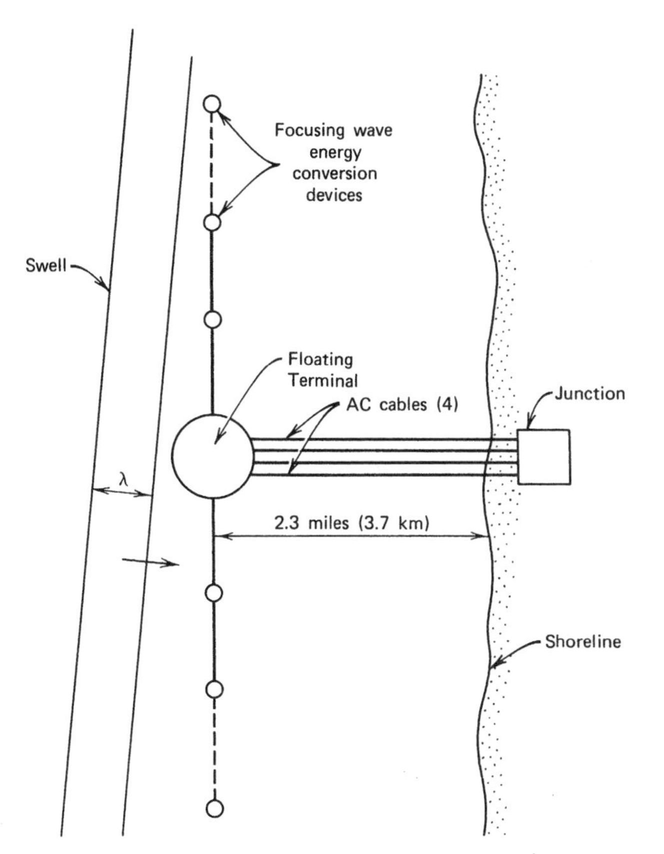

Figure 5.14 Schematic diagram of cable transmission of converted wave energy of Example 5.7.

terminal should be dc for reasons presented in Sections 5.1 and 5.2. Thus a dc–ac convertor is required on the terminal.

For swell that have a linear power intensity of 10 kW per foot of crest (32.8 kW/mr), the array of devices would span 7.58 miles (12.2 km) parallel to the shoreline if operating at 25% efficiency to deliver 100 MW to the terminal.

Direct cable transmission has been successfully tried in the Kaimei test in the Sea of Japan (discussed in Section 4.1, B). Preliminary results of this test are presented by Miyasaki and Masuda (1980).

B Energy-Intensive Products

When the distance between a wave energy conversion system and the shore exceeds 80 miles (129 km), direct electrical transmission by cable is not feasible. This situation would be enountered if large-scale wave energy conversion occurs in the trade winds belts. As in the case of direct electrical energy transmission, much effort connected with OTEC has been devoted to alternative energy transmission and storage methods. These methods involve the manufacturing of energy-intensive products such as aluminum. Since the wave resource is relatively variable when compared to the ocean thermal resource, care must be taken as to the type of product manufactured.

Winer (1975) recommends consideration of producing *metals* that require much electrical energy in their production, such as aluminum. He also suggests the manufacture of *synthetic fuels*. In both cases some *raw materials* would be transported to the energy conversion site from land. In the case of aluminum production, bauxite would be required, whereas coal would be required in coal gasification production. Konopka et al. (1977) further consider the production of fuels of the *hydrazine* family and more conventional hydrocarbon fuels in connection with OTEC. The findings of Konopka et al. (1977) indicate that the delivered costs of these manufactured fuels are much higher than those of *liquefied hydrogen* and *liquefied ammonia*, which can also be manufactured on site without requiring the transport of raw materials. Further studies are being performed in these areas in connection with OTEC.

Both Winer (1975) and Konopka et al. (1977) suggest using large *lithium batteries* for the transport of dc electricity over long distances. Konopka et al. (1977) perform an in-depth analysis of this energy transmission and storage technique, which they refer to as the *electrochemical bridge*. Although the development of this large battery system is somewhat in the future, this author finds the lithium battery barge apropos for wave energy conversion because of its *simplicity* and relatively *high transmission efficiency*.

An excellent description of the *lithium–water—air battery* is given by Konopka et al. (1977). Referring to the scheme shown in Figure 5.15, a barge transports a lithium hydroxide (LiOH) slurry to the wave energy conversion terminal where, by using the wave energy converted to electricity, the slurry is reduced to molten lithium metal, oxygen, and water. The lithium metal is

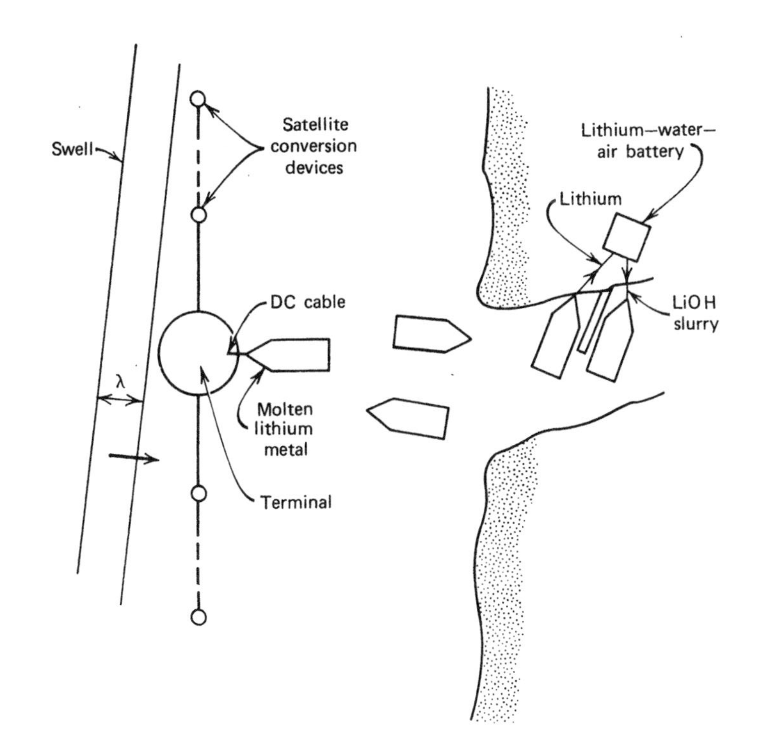

Figure 5.15 Possible scheme for transmission and storage of converted wave energy using the "electrochemical bridge" of Konopka et al. (1977).

transported to shore where the lithium–water–air battery uses the lithium as its anode, which is consumed during discharge. Lithium hydroxide produced during the discharge of the battery is then recovered by cooling the electrolyte for transport back to the wave energy conversion terminal.

Example 5.8

Konopka et al. (1977) give several examples of applications of energy-intensive products to OTEC systems. One such example involved the electrochemical bridge. For a 100-MW plant reducing 180.36 tons of lithium per day, 62.1 MW of energy is delivered from the onshore battery. Thus the overall efficiency of the energy transfer system is 62.1%.

5.5 Summary

From the discussions presented in Chapter 4, the reader can conclude that certain wave energy conversion techniques have excellent efficiencies. The materials presented in Chapter 5, however show that the overall efficiency is determined by the electrical energy conversion technique and the method of energy transfer.

References

Anonymous (1978), "Electrical Generator Uses Ocean Waves," *Mechanical Engineering*, June, p. 53.

Babinsten, I. A. (1975), "Apparatus for Converting Sea Wave Energy into Electrical Energy," U.S. Patent No. 3,922,729, December 2.

Bamford, T., Dow, H., Libby, D., Munn, P., Pieroni, C., and Traut, R. (1978), "Riser Segment Design of Underwater Electric Power Transmission Cable System," U.S. Department of Energy Report ORO/5359-1, October.

Bhattacharyya, R. (1978), *Dynamics of Marine Vehicles*, Wiley-Interscience, New York.

Binns, K. J. (1979), "The Use of Permanent Magnet Machines for Low Speed Generation," *Proceedings, Wave Energy Utilization Symposium*, Gotheburg, Sweden.

Bishop, H. W., and Rees, G. R. (1979), "An Electrical Generation and Transmission Scheme for Wave Power," *Proceedings, Wave Energy Utilization Symposium*, Gotheburg, Sweden.

Burfoot, J. C., and Taylor, G. W. (1979), *Polar Dieletrics and Their Applications*, Macmillan, London.

Filipenco, G. D. (1975), "Electrical Stations Operated by Waves," U.S. Patent No. 3,912,932, October 14.

Garrity, T. F., and Morello, A. (1979), "A Theoretical Study of Technical and Economic Feasibility of Bottom Submarine Cables for OTEC Plants, "*Proceedings, 6th OTEC Conference*, Washington, D.C., June, Paper 7.1.

Hiramoto, A. (1978), "The Theoretical Analysis of an Air Turbine Generation System," *Proceedings, Wave and Tidal Energy Symposium*, Canterbury, England, Paper B5.

Konopka, A., Talib, A., Yudow, B., Blazek, C., and Biederman, R. (1977), "Alternative Energy Transmission Systems from OTEC Plants," U.S. Department of Energy Report No. DSE/2426-20.

Masuda, Y., and Miyazaki, T. (1978), "Wave Power Electric on Study in Japan," *Proceedings, Wave and Tidal Energy Symposium*, Canterbury, England, Paper B6.

McCormick, M. E. (1973), *Ocean Engineering Wave Mechanics*, Wiley-Interscience, New York.

McCormick, M. E. (1974) "An Analysis of a Stationary Wave-Energy Converter," American Society of Mechanical Engineers, *Paper No. 740WA/OCT-2*, October.

McCormick, M. E. (1978), "Ocean Wave Energy Conversion Concepts," *Proceedings, Oceans '79 Conference*, San Diego, Calffornia (MTS-IEEE), pp. 553–557.

Miyazaki, T., and Masuda, Y. (1980), "Development of Wave Power Generator Buoy, Kaimei," *Proceedings, Offshore Technology Conference*, Houston, Texas, May, Paper OTC 3689.

Oliver, J. C., and Jawish, W. K. (1979), "Load Criteria for OTEC Riser Cable Design," *Proceedings, 6th OTEC Conference*, Wawhington, D.C., June, Paper 7.3.

Omholt, T. (1978), "A Wave Activated Electric Generator," *Proceedings, "Ocean '78," Marine Technology Conference*, Washington, D.C., pp. 585–589.

Pieroni, R., Traut, R., Libby, D., and Garrity, T. (1979), "The Development of Riser Cable Systems for OTEC Plants," *Proceedings, 6th OTEC Conference*, Washington, D.C., June, Paper 7.2.

Salomon, R. E. (1978), "Protonic Conduction Wave Energy Convertor," U.S. Department of Energy Report No. ORO-5669-T1.

Salomon, R. E., and Harding, S. M. (1979), "Gas Concentration Cells for the Conversion of Ocean Wave Energy," *Ocean Engineering*, Vol. 6, No. 3, pp. 317–327.

Taylor, G. W. (1979), "Piezoelectric Power Generation from Ocean Waves," Princeton Resources, Princeton, N.J., report, March.

Thorborg, K. (1979), "Frequency Converter for Ocean Wave Utilization," *Proceedings*, *Wave Energy Symposium*, Gothenburg, Sweden.

Whittaker, T. J. T., and Wells, A. A. (1978), "Experiences with Hydropneumatic Wave Power Device," *Proceedings*, *Wave and Tidal Energy Symposium*, Canterbury, England, September, Paper B4.

Wilson, M. N. (1979), "Slow Speed Generators with Superconducting Windows," *Proceedings*, *Wave Energy Utilization Symposium*, Gothenburg, Sweden.

Winer, B. M. (1975), "Electrical Energy Transmission from Ocean Thermal Power Plants," *Proceedings*, *Third Workshop on Ocean Thermal Energy Conversion*, Houston, Texas, May, pp. 103–105.

6 Environmental and Mooring Considerations

This last chapter deals with two important aspects of wave energy conversion: (*a*) environmental considerations and (*b*) mooring and anchoring of wave energy conversion systems. The topics are grouped together simply because they can be considered to be *ancillary*. There are several references that collectively cover these topics in great detail. For example, Dawson's (1979) report discusses both the environmental issues and the technical and economic aspects of mooring and anchoring, among other topics related to the wave energy program in the United Kingdom. Much information on the topic areas has also been obtained in connection with the OTEC program in the United States. This information is presented in the proceedings of the various OTEC conferences. For example, see those proceedings edited by Dugger (1979).

The application of each of the topics of this chapter depends on both the *location* and the *type* of wave energy conversion system. Problems associated with deep water operations in the open ocean are far different from those encountered in the coastal zone. For this reason the sections that follow are each divided into two subsections dealing with *open ocean operation* and *coastal zone operation*.

6.1 Environmental Considerations

The environmental aspects of wave energy conversion include the effects on *coastal processes*, *wildlife*, and *aesthetics*. These effects need not be damaging, but may in some instances be beneficial. Excellent discussions of the environmental aspects are found in the publications of Dawson (1979), Lewis and Waite (1980), and the Environmental Analysis Corporation (Anonymous, 1979).

A Open Ocean Operation

Wave energy conversion in the deep waters of the open ocean has the least number of environmental problems. The conversion devices are located far from the shoreline so that they do not affect beach stability. Furthermore, the devices cannot be seen from the shore and present no aesthetic problems. The single environmental issue that is often raised concerning open ocean wave energy conversion concerns the effect on sealife. Wave action helps support life in the surface waters by *circulating oxygen* and *nutrients*.

In the *trade wind belts* the candidate waves for energy conversion are generated by winds that are nearly constant in both speed and direction, as discussed in Section 2.3. When part of the wind-wave energy is converted, a generating distance (called a *fetch*) is required for the wind to restore the wave energy to the fully developed value. A *fully developed sea* is one in which the energy (or power) spectra are constant, as discussed in Section 2.3. Wind-wave generation is thoroughly discussed in the publications of the U.S. Army (1973), Sorensen (1978), and Kinsman (1965) (the latter is a classic).

Consider the situation sketched in Figure 6.1, where several wave energy conversion systems in a constant wind region are sketched. The systems are similar to that sketched in Figure 5.14. For a constant wind speed V, a fully developed sea with an energy spectrum similar to that shown in Figure 2.8 will occur. This condition is assumed to exist up wind of system A in Figure 6.1. Assuming a wave energy *conversion efficiency* ϵ, a portion of the *upwind wave power* P_0 is absorbed by the system, and the remaining portion is transmitted. The *average power transmitted* is obtained from

$$P_1 = P_0(1 - \epsilon) \tag{6.1}$$

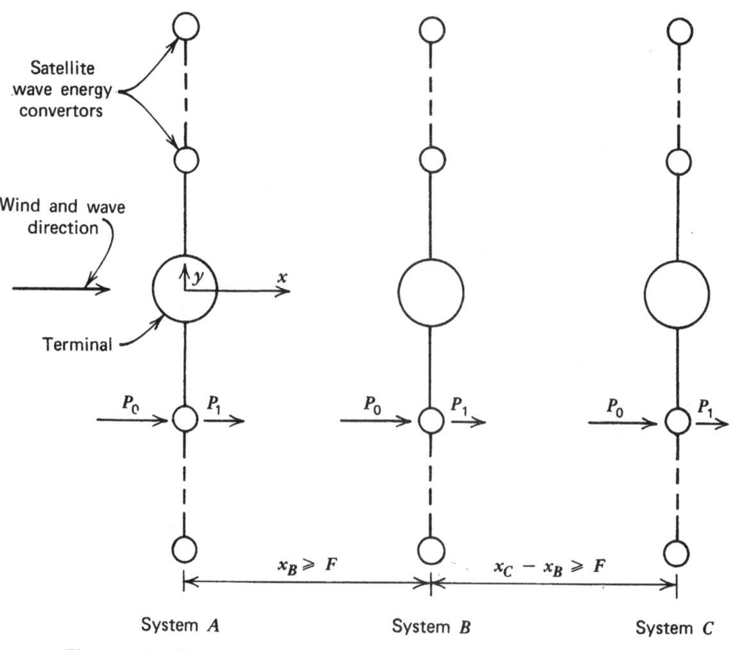

Figure 6.1 Plan view sketch of three wave energy conversion systems.

Since the wind speed is constant in both magnitude and direction, we can locate system B downwind from system A such that the sea available to system B will again be fully developed. Thus the distance between systems A and B (also B and C) must be at least equal to the *minimum fetch* F, that is, the fetch required to generate a fully developed sea.

The prediction of *wind-wave properties* as functions of *wind speed, fetch,* and *duration* is accomplished by using the curves in Figure 6.2, taken from the U.S. Army (1973) publication. The analysis leading to these curves was performed by Bretschneider. See Bretschneider (1967). In Figure 6.2 the wave properties are presented in terms of the *significant wave height* H_s and the *significant wave period* T_s discussed in Section 2.3. The significant wave height is the average of the one-third highest waves, whereas the significant wave period is the average period of these waves. For all practical purposes, the significant wave period and *average wave period* $\overline{T}$ of equation (2.46) are the same.

There is *one basic difference* in the wind-wave analysis of Section 2.3 and that of Bretschneider leading to the curves shown in Figure 6.2, specifically, in the *measurement of the wind speed.* The wind speed in Section 2.3 is measured at 64.0 ft (19.5 m) above the SWL and is identified by V. The wind speed in Figure 6.2 is measured at a point 32.8 ft (10.0 m) above the SWL and is

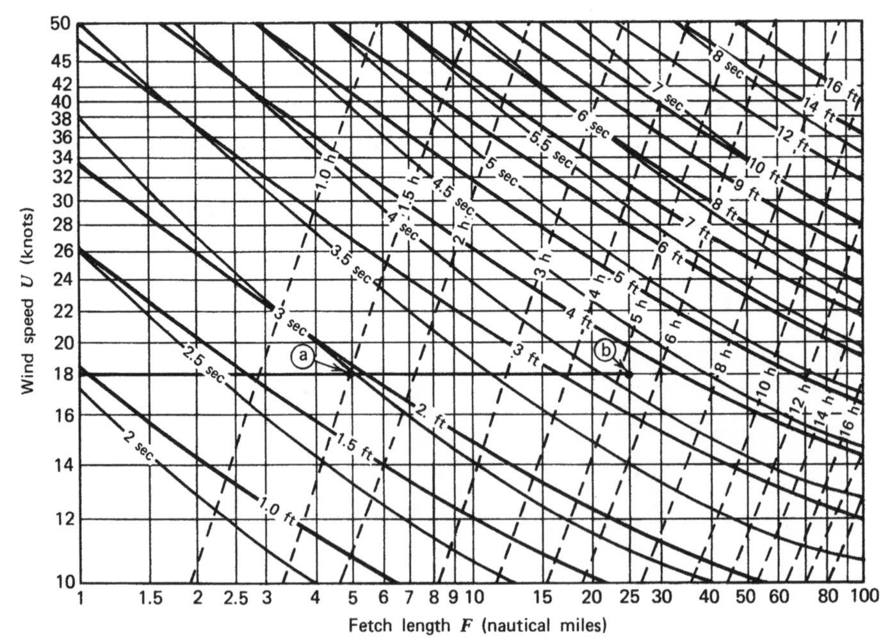

Figure 6.2 Deep water wave forecasting curves, where heavy diagonal lines represent H_s, soft diagonal lines represent T_s, and dashed lines represent wind duration. From U.S. Army publication (1973).

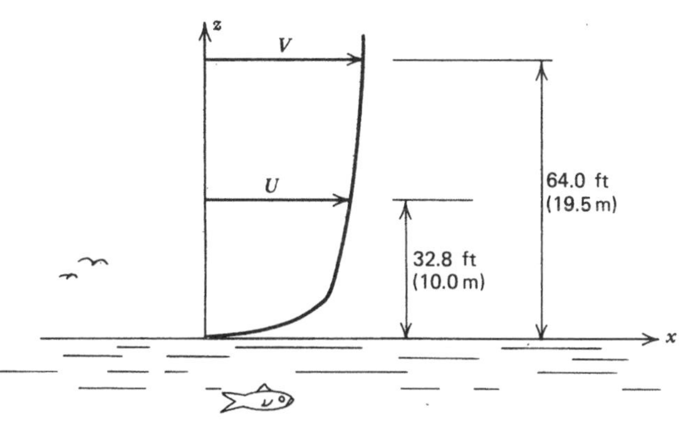

Figure 6.3 Sketch illustrating the air velocity profile over calm water, where U is the velocity assumed in Figure 6.2 and V is that assumed in Figure 2.8.

identified by U. Assuming a *one-seventh power law* for the wind velocity distribution over the water (Streeter, 1971), the relationship between the two wind speeds is

$$U = 0.909 V \tag{6.2}$$

where, again, U is measured at $z = 32.8$ ft (10.0 m) and V is measured at $z = 64.0$ ft (19.6 m). The difference in these wind speeds can be seen in Figure 6.3.

The use of the curves in Figure 6.2 is illustrated in Example 6.1.

Example 6.1

At a site in the Northeast Trades (see Figure 1.2) the prevailing wind speed measured at $z = 64.0$ ft (19.5 m) is 20 knots (33.8 ft/sec or 10.3 m/sec). There is little variation in either the speed or the direction of the wind at this location, and thus the *wind duration* can be considered to be *infinite*. The sea is, therefore, *fully developed*, which means that the statistical wave properties H_s and T_s and the total wave energy are constant in value. For the 20-knot wind speed, the *peak energy* occurs at a wave period of 6.60 sec, from both Figure 2.8 and Example 2.9. The significant wave properties obtained in Example 2.10 are

$$H_s = 7.41 \text{ ft } (2.26 \text{ m})$$

and

$$T_s \simeq \overline{T} = 6.60 \text{ sec}$$

The average wave power per crest width b corresponding to these values is

$$\frac{P_0}{b} = \frac{\rho g^2 H_s^2 T_s}{64\pi}$$
$$= \frac{2.00(32.2)^2(7.41)^2(6.60)}{64\pi}$$
$$= 3740 \text{ lb-ft/s-ft } (16.6 \text{ kW/m}) \qquad (6.3)$$

Since we are concerned with the effect of wave power conversion on mixing, we shall take the most extreme case. That is the case of 100%-efficient wave energy conversion devices in systems A, B, and C sketched in Figure 6.1. Thus, from equation (6.1), where $\epsilon = 1.00$, the transmitted power P_1 is zero. A fetch F is required for the wind to replenish the power to the sea.

Actually, the power of a fully developed sea will not be resupplied by the 20-knot wind over any reasonable length. Wind-wave generation will begin in the lee of system A, and the wave power at a given distance downwind of the system is found by entering the left-hand side of Figure 6.2 with the wind velocity measured at $z = 32.8$ ft (10.0 m), that is, from equation (6.2)

$$U = 0.909 \text{ V}$$
$$= 0.909(20)$$
$$= 18.2 \text{ knots } (30.7 \text{ ft/sec or } 9.36 \text{ m/sec})$$

A horizontal line drawn to the right of the U-axis at the value of U just given passes through lines corresponding to H_s and T_s or $\overline{T}$ at any fetch value. For example, at point a in Figure 6.2,

$$F \simeq 5.00 \text{ nautical miles } (5.76 \text{ statute miles or } 9.27 \text{ km})$$
$$H_s \simeq 1.92 \text{ ft } (0.586 \text{ m})$$
$$T_s \simeq 2.96 \text{ sec}$$

The power per crest length at point a, from the results in equation (6.3), is

$$\frac{P_a}{b} = \frac{\rho g^2 H_s^2 T_s}{64\pi}$$
$$= \frac{2.00(32.2)^2(1.92)^2(2.96)}{64\pi}$$
$$= 112 \text{ lb-ft/sec-ft } (0.500 \text{ kW/m})$$

Similarly, at point b in Figure 6.2,

$$F = 25.0 \text{ nautical miles (28.8 statute miles or 46.3 km)}$$

$$H_s \simeq 3.58 \text{ ft (1.09 m)}$$

$$T_s \simeq 4.41 \text{ sec}$$

$$\frac{P_b}{b} = 583 \text{ lb-ft/sec-ft (2.59 kW/m)}$$

The reader can see that the wave power at point b in Example 6.1 is less than 20% of the fully developed sea power. For this reason, use of 100% efficient wave energy conversion devices in long arrays may pose serious environmental problems.

From the discussions in Chapter 4, the most effective wave energy conversion devices are *frequency dependent*. These devices, such as the resonant heaving body described in Section 4.1, A or the cavity resonator described in Section 4.1, B, can take advantage of "*antenna*" *focusing* (see beginning of Section 4.2, D1) because of their resonant nature. Since these devices are *tuned* to a specific wave frequency, they are most effective in converting the energy of waves that have this specific frequency. Thus waves that have frequencies outside the neighborhood of the resonant frequency will be little affected by the device. For this reason the energy spectrum in the lee of the wave energy conversion system will be modified but will not vanish. Significant mixing will still occur; therefore, *wave energy conversion in the open ocean does not present an environmental problem if frequency-dependent wave energy conversion devices are used.*

B Coastal Zone Operation

Wave energy conversion near the coast will present an aesthetic problem since the devices will be visible from the shore. A line of Salter's nodding ducks (Section 4.2, A) several miles in length probably would be more aesthetically unacceptable than a series of well-spaced "point" absorbers (see beginning of Section 4.2, D). Furthermore, a long continuous line of wave energy conversion devices would be an *impediment to navigation*, whereas point absorbers would not hamper navigation.

The most important environmental consideration in coastal zone wave energy conversion concerns the effects on the *littoral processes*. The materials moved by waves in the coastal zone are called *littoral drift*. The movement of these materials is necessary for *beach stability*, that is, the balance between *erosion* and *accretion*. When a region of the coast is totally deprived of wave energy by an offshore *breakwater*, a *tombolo* will form, as illustrated in Figure 6.4. A tombolo is a *sand bar* extending from the shoreline to the breakwater.

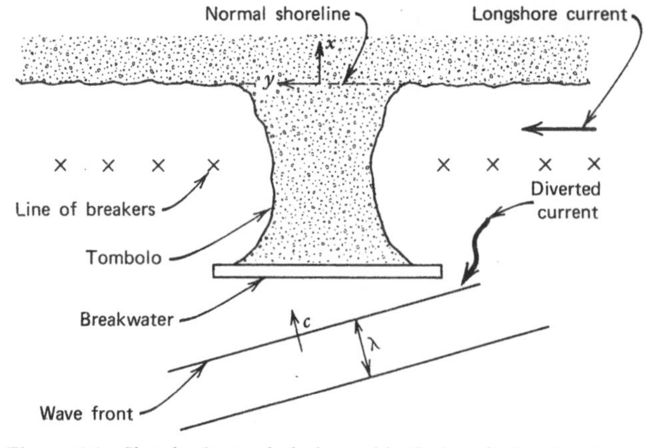

Figure 6.4 Sketch of a tombolo formed in the lee of a breakwater.

When a tombolo is formed, the longshore current carrying the littoral drift is diverted offshore. The result may be a total deprivation of sand to the downdrift region of beach. Obviously, a line of 100%-efficient wave energy convertors will have the same effect as that of a breakwater. For this reason a number of engineers have suggested using wave energy conversion devices in place of breakwaters to perform the *dual functions* of wave energy conversion and coastal protection. This would improve the cost-effectiveness of the wave energy conversion system.

The longshore current within the surf zone is caused by wave action. The wave power at the line of breakers P_b has a longshore component P_l, which drives this current. Referring to Figure 6.5, the magnitude of P_l depends on the deep water wave angle β_0, deep water wave height H_0, period T, and

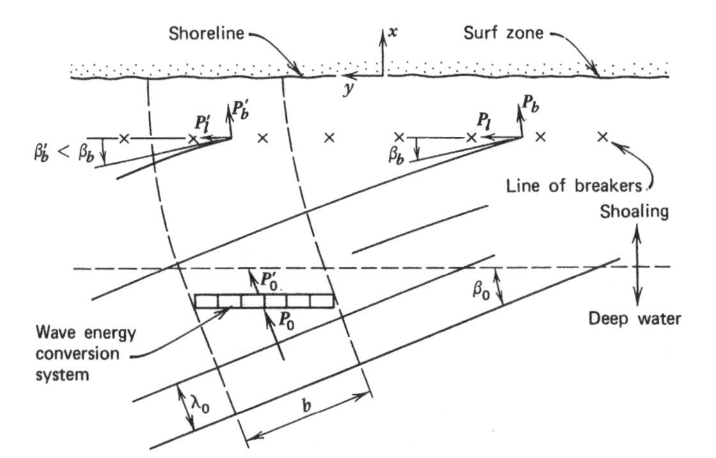

Figure 6.5 Plan view sketch of the effect of a wave energy conversion system on the longshore power in the surf zone.

breaker angle β_b. The relationships among these various wave properties are presented in Figure 6.6, which is taken from the U.S. Army (1973) publication. To gain an understanding of the possible effect of wave energy conversion on the littoral or longshore current, consider Example 6.2.

Example 6.2

A deep water swell that has a height H_0 of 3 ft (0.914 m) and period T of 6 secs approaches a coast at 30° to the shoreline, that is,

$$\beta_0 = 30°$$

in Figures 6.5 and 6.6. A line of wave energy convertors is situated parallel to the coastline in deep water and converts the power of the swell with an efficiency, ϵ, of 50%. The wave power available to the surf zone is then

$$\begin{aligned} P_0' &= P_0 - \epsilon P_0 \\ &= P_0(1 - 0.5) \\ &= 0.50 P_0 \end{aligned} \tag{6.4}$$

The expression for the deep water wave power is obtained from equation (2.19) by using the results of equations (2.20b) and (2.7). The result is

$$P_0 = \frac{\rho g^2 H_0^2 T b}{32\pi} \tag{6.5}$$

where b is the width of the affected wave front. The conversion of the wave power changes the wave height but not the wave period. Thus on the lee side of the wave energy conversion system, the wave power is

$$P_0' = \frac{\rho g^2 (H_0')^2 T b}{32\pi} = \frac{(1 - \epsilon)\rho g^2 H_0^2 T b}{32\pi} \tag{6.6}$$

The wave height in the lee is then

$$\begin{aligned} H_0' &= \sqrt{1 - \epsilon}\, H_0 \\ &= \sqrt{1 - 0.50}\,(3) \\ &= 2.12 \text{ ft } (0.646 \text{ m}) \end{aligned}$$

The swell not exposed to the wave energy conversion breaks with an angle of

$$\beta_b = 10.5°$$

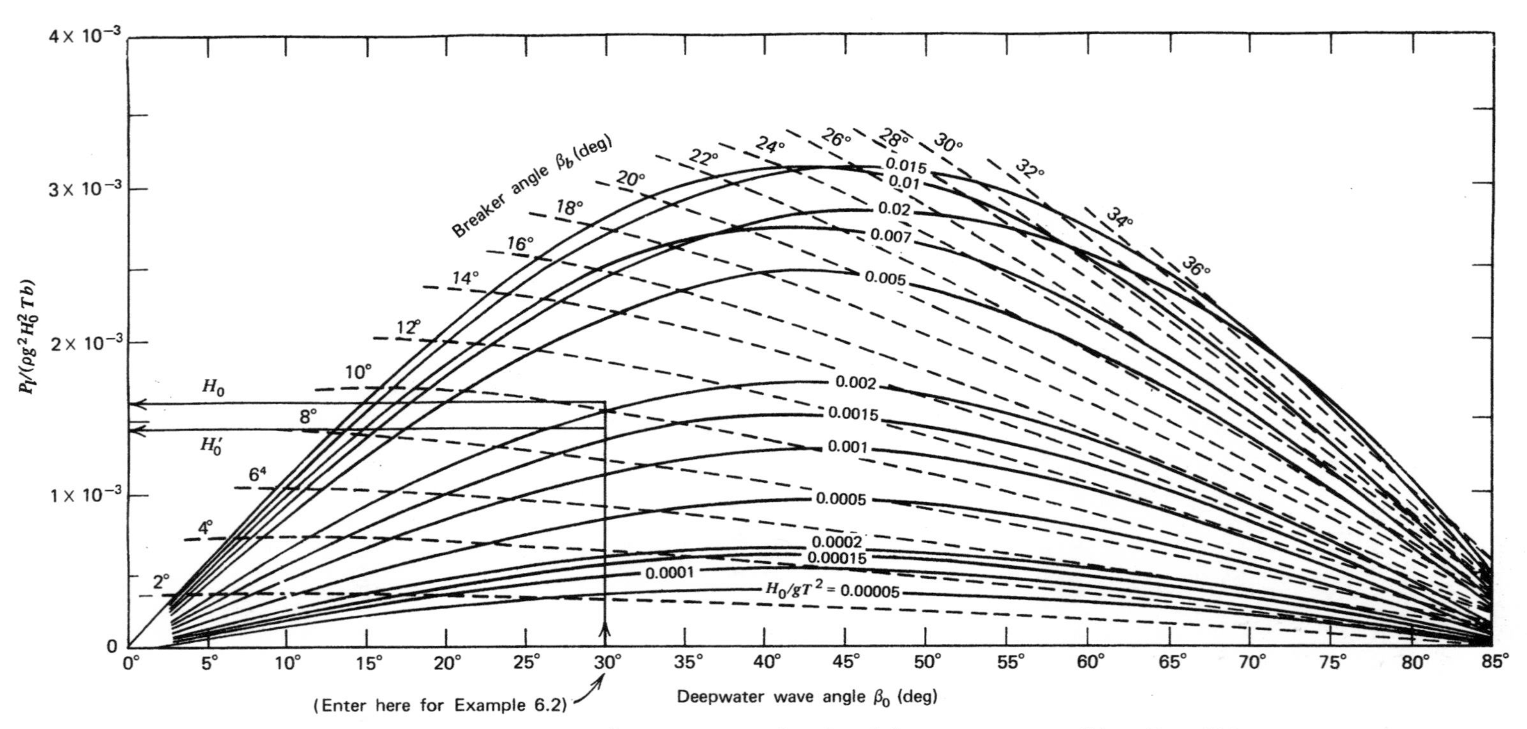

Figure 6.6 Longshore component of wave power as a function of deep water wave conditions. From U.S. Army (1973) publication.

to the shoreline, from the results of Figure 6.6. This value of β_b is obtained by drawing a vertical line from the deep water wave angle axis at

$$\beta_0 = 30°$$

and finding that line's intersection with the line (interpolated) corresponding to the *deep water parameter*

$$\frac{H_o}{gT^2} = \frac{3}{32.2(6)^2}$$

$$= 2.59 \times 10^{-3}$$

By drawing a horizontal line from the point of intersection to the left-hand axis, the value of the longshore *wave power parameter* is obtained; thus

$$\frac{P_l}{\rho g^2 H_0^2 T b} \simeq 1.56 \times 10^{-3}$$

so that the *longshore power per crest length* is

$$\frac{P_l}{b} = 1.56 \times 10^{-3}(\rho g^2 H_0^2 T)$$

$$= 1.56 \times 10^{-3}\left[2.00(32.2)^2(3)^2 6\right]$$

$$= 175 \text{ lb-ft/sec-ft } (0.778 \text{ kW/m})$$

With the wave energy conversion system on site, the *effective deep water wave height* H_0 is 2.12 ft (0.646 m), so that

$$\frac{H_0'}{gT^2} = \frac{2.12}{32.2(6)^2}$$

$$= 1.83 \times 10^{-3}$$

The vertical line from $\beta_0 = 30°$ in Figure 6.6 now intersects the interpolated line corresponding to $H_0'/gT^2 = 1.83 \times 10^{-3}$ at a breaker angle of 9.5°. The corresponding longshore power parameter is

$$\frac{P_l'}{\left(\rho g^2 (H_0')^2 T b\right)} = 1.40 \times 10^{-3}$$

so that the new longshore power per length is

$$\frac{P_l'}{b} = 1.40 \times 10^{-3}(\rho g^2 H_0'^2 T)$$

$$= 1.40 \times 10^{-3}\left[2.00(32.2)^2(2.12)^2 6\right]$$

$$= 78.3 \text{ lb-ft/sec-ft } (0.348 \text{ kW/m})$$

From the results of Example 6.2 we see that the extraction of offshore wave power reduces both the breaker angle and the longshore wave power. The reduction in the longshore wave power is more than one would expect. If we convert 50% of the deep water wave power, then logic would dictate that the value of P_l is similarly reduced by 50%. However, β_b is also reduced, so the new value of the longshore current is less than 50% of P_l. Since P_l is the power moving the littoral drift, the longshore volume rate of flow within the surf zone is also reduced. This reduction will result in an accretion of materials that may be beneficial if the beach has been unstable, that is, having a net *erosion*. If wildlife breeding areas are in the region, then the effect of the wave energy conversion could be damaging since the accretion could cover breeding beds or deprive a saltwater march of flushing currents.

6.2 Mooring and Anchoring

Ocean engineers concerned with the design of mooring systems for offshore structures have a goal of *minimizing* wave-, current-, and wind-induced motions of these structures. Thus it would be contrary to these engineers' experiences to design mooring systems for wave energy conversion devices that undergo large excursions (by design). If the motions associated with wave energy conversion can be *isolated* from the mooring system, as is the case illustrated in Figure 4.7, the *efficiency* of the wave energy conversion device is maximized. This fact is well illustrated by the results of Carmichael (1978), shown in Figure 4.26. The type of mooring depends on the nature of the wave energy conversion device, depth of water, environment (wind, waves, and current), tidal range, method of deployment, and designed life of the device. The mooring of a wave energy conversion system is a major part of the system design. The costs of the moorings can exceed the other subsystem costs and, therefore, be the determining factor in the cost-effectiveness of the total system.

The publication edited by Dawson (1979) contains an excellent section on mooring and anchoring of wave energy conversion systems. The reader is also referred to the studies involving OTEC that include the work of the Frederic R. Harris Corporation (Anonymous, 1977) and that of Valent and Atturio (1977). The latter half of the book by Berteaux (1976) contains one of the best detailed discussions of mooring and anchoring available.

A Deep Ocean Operation

The first step in the design of a mooring system is the determination of the *configuration*. In deep water the candidates are a *slack mooring* using a *multicomponent line*, as discussed by Ansari (1979), and the *tension mooring*, as discussed by Hong (1974). For both of these configurations some type of *embedment anchor* is advised (Berteaux 1976; Valent and Atturio, 1977). In waters of extreme depth a *sea anchor* may also be suitable. The sea anchor can be used in conjunction with *dynamic positioning*, as studied by the Frederic R. Harris Corporation (Anonymous, 1977). Dynamic positioning requires power, and this fact must be weighed against the costs of a fixed mooring.

A slack mooring configuration was used in the full-scale wave energy conversion tests conducted on the Kaimei in the Sea of Japan (Miyazaki and Masuda, 1980; McCormick and Masuda, 1980). A heavy-chain mooring was used in that study, a link of which is shown in Plate 6.1. A plan view of the mooring system of the Kaimei is sketched in Figure 6.7*a*. The water depth at the site of the Kaimei tests is 130 ft (40 m) and can be considered as intermediate.

The most *practical* deep ocean system configuration is that sketched in Figure 5.15, where a central terminal receives the converted wave energy from an array of wave energy conversion devices. In this case the terminal would receive the most attention as far as mooring is concerned. Since minimum motions of the terminal are desired, the terminal itself should be a *spar buoy* with a relatively small waterplane area. From the results of equation (4.3), the *waterplane area* A_{wp} affects the *heaving frequency* f_z. The spar configuration, sketched in Figure 6.8, yields large values of the *mass moment of inertia* I_y and

Plate 6.1 The 95-mm mooring chain used on the Kaimei in the Sea of Japan. Courtesy of the Japan Marine Science and Technology Center.

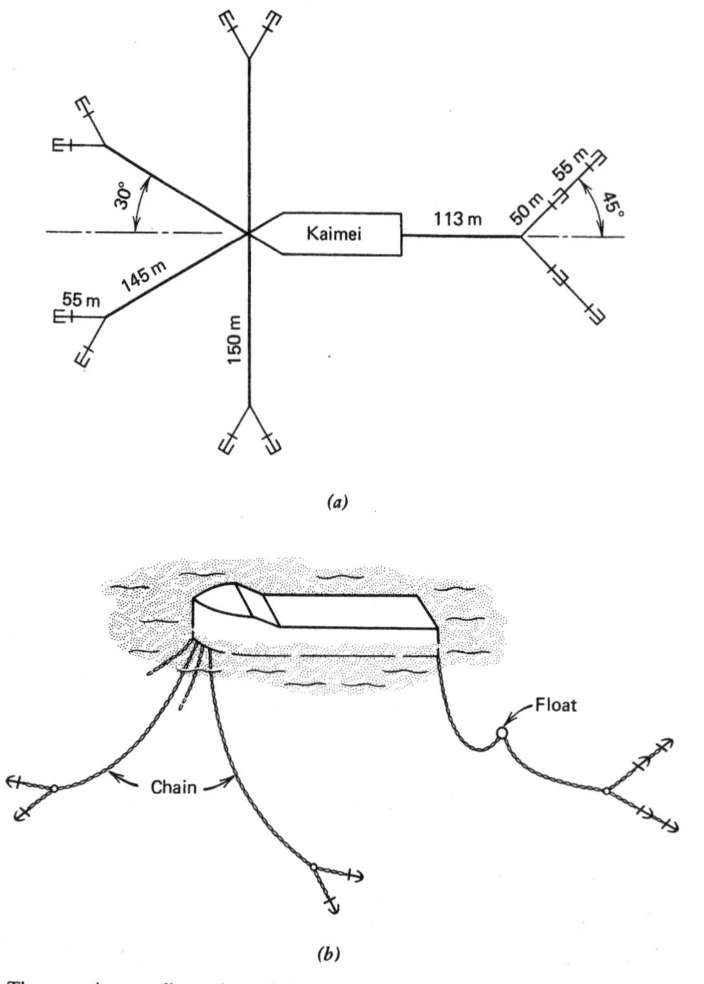

Figure 6.7 The mooring configuration of the Kaimei in the Sea of Japan: (*a*) plan view; (*b*) sketch.

the *added-mass moment of inertia* I_w, both of which affect the *pitching frequency* f_θ given in equation (4.6). Either extremely high or low values of both f_z and f_θ are desired so that these natural frequencies be well away from the high-energy waves. For example, for the wave energy spectrum shown in Figure 2.8, heaving and pitching wave periods greater than 12 secs would be desired. These periods correspond to frequency values of 0.0833 Hz, since $f = 1/T$. Since the wave-induced motions are minimal for a spar buoy, a *single-point slack mooring* with a *swivel connection* to the buoy is apropos, as sketched in Figure 6.8. This configuration allows for a certain amount of drift resulting from the wind loading on the superstructure and the wave loading on the hull.

If a spar buoy is impractical for the terminal, and a *large waterplane area* is

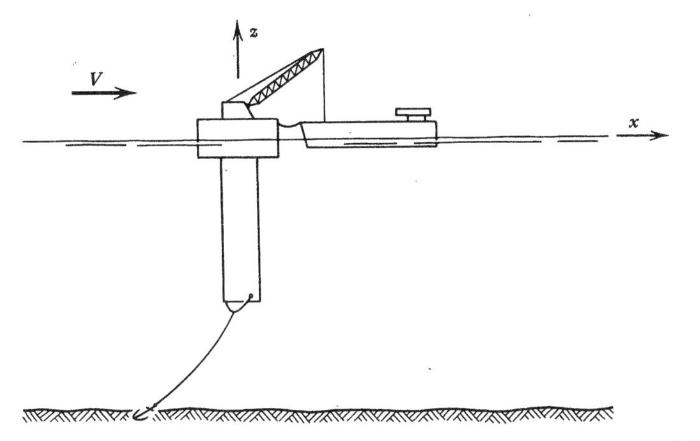

Figure 6.8 Wave energy conversion terminal with a semi-spar hull form to minimize wave-induced motions. The terminal is slack moored.

needed, a *tension mooring* would be necessary to reduce the wave-induced heaving and pitching motions. In this case the expressions for the heaving frequency f_z and the pitching frequency f_θ must be modified to include the "*spring*" *effect* of the mooring lines. Referring to Figure 6.9, equations (4.3) and (4.6), respectively, become

$$f_z = \frac{1}{T_z} = \frac{1}{2\pi}\sqrt{\frac{\rho g A_{wp} + nk_c}{m + m_w}} \tag{6.8}$$

and

$$f_\theta = \frac{1}{T_\theta} = \frac{1}{2\pi}\sqrt{\frac{C + Dk_c(n/2)}{I_y + I_w}} \tag{6.9}$$

from McCormick (1973). In equations (6.8) and (6.9) n is the total number of mooring lines, each of which has a *spring constant* of k_c, m_w is the added mass, I_y is the mass moment of inertia of the terminal with respect to the y-axis, and I_w is the added-mass moment of inertia with respect to the y-axis. For expressions of m_w and I_w, see Figures 4.2, 4.3, and 4.4. Assuming the terminal to have a circular waterplane area of diameter D, the waterplane expression is

$$A_{wp} = \frac{\pi D^2}{4} \tag{6.10}$$

The results in equation (6.8) show that the heaving frequency for a specified terminal can be increased either by increasing the number of mooring lines n or by altering the value of the spring constant k_c. The

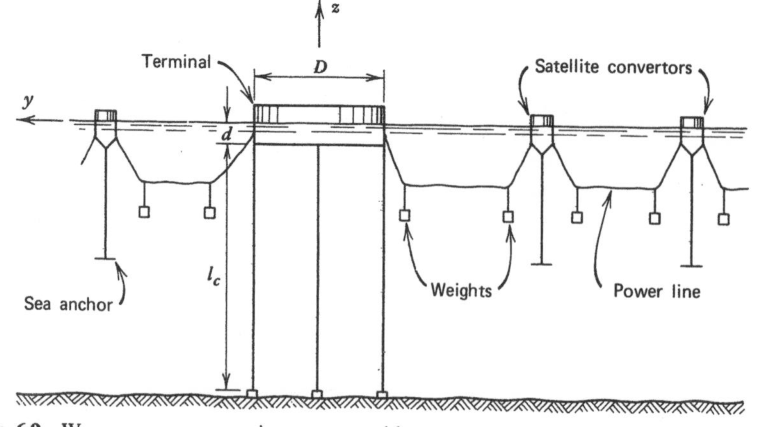

Figure 6.9 Wave energy conversion system with a taut moored terminal and sea-anchored convertors.

expression for the spring constant in terms of the cable properties is

$$k_c = \frac{Y_c \pi r_c^2}{l_c} \tag{6.11}$$

where Y_c is the modulus of elasticity of the line material, r_c is the line radius, and l_c is the unstretched length of the line. Thus the most practical way to increase the value of k_c (and, thereby, both f_z and f_θ) is to decrease l_c.

The expression for the pitching natural frequency in equation (6.9) assumes a four-point mooring configuration where, for the open ocean in a region where the wind and wave directions are relatively constant, a *restoring couple* of $DK_c n/2$ results from the "bow" and "stern" lines. The terms "bow" and "stern" refer, respectively, to the upwind and downwind lines. The two lines "amidships" do not contribute to the restoring moment. This configuration is rather ideal in the sense that the wind and wave directions do change. Furthermore, more than four anchor points might be desirable.

The mooring of the *satellite wave energy conversion devices* sketched in Figure 6.9 poses a more difficult problem since these devices must be kept well away from each other. This is assuming that the devices are resonant in nature and can take advantage *antenna focusing*, as discussed in the beginning of Section 4.2, D. The satellite devices are much smaller than the terminal and hence require less massive mooring systems. For a deep water site in the trade winds, a slack mooring attached to a *sea anchor* would suffice to keep the device erect. Remotely controlled *thrusters* could be used in conjunction with the mooring. The power for the thrusters could be supplied by a *windmill–battery system* on each device or by using some of the converted wave energy. A sketch of a wind-driven system is presented in Figure 6.10. The system shown in Figure 6.10 consists of a pneumatic wave energy convertor with power lines connecting the device to other satellites and the terminal. The

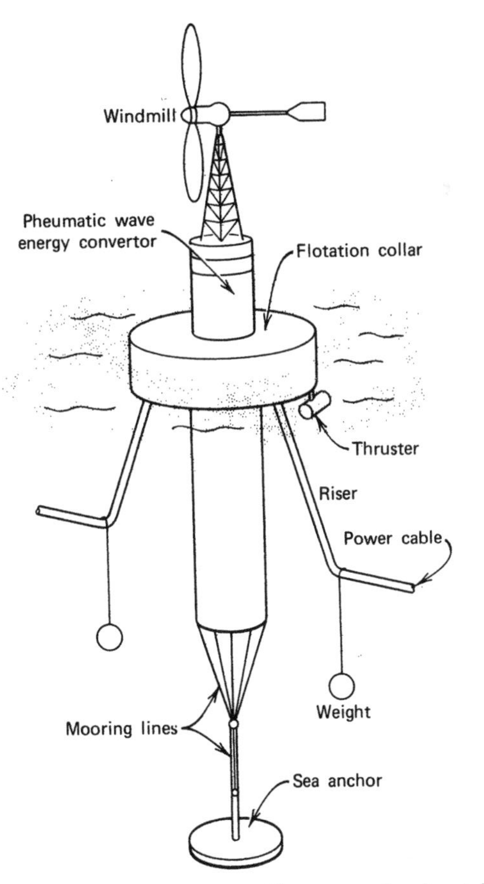

Figure 6.10 A satellite wave energy convertor with a sea anchor for stability and a windmill-powered thruster for station keeping.

power lines, attached to the device by using *risers*, are kept well below the free surface by weights. This is necessary for the lines to be relatively free from both storm conditions and ships.

The materials used in mooring lines are thoroughly discussed by Berteaux (1976). The materials fall into three categories: (*a*) *metals* (wire or chain), (*b*) *synthetic materials* (nylon line, etc.), and (*c*) *natural nonmetallics* (manila rope, etc.). Of these three, only categories *a* and *b* are considered to be practical for wave energy conversion systems. A summary of material properties is presented in the publication edited by Dawson (1979). These properties are shown in Table 6.1. The relative costs of the synthetic lines is discussed by Heller (1970), although inflation since 1970 has significantly changed the actual costs.

Berteaux (1976) presents a table of performance values for synthetic lines that the reader is urged to consult. From the discussion of Berteaux (1976), the most appropriate candidates for deep water moorings of wave energy convertors are those described in the following two paragraphs.

Table 6.1 Mooring Material Properties

Material	Corrosion Resistance	Fouling Resistance	Weight	Elasticity	Buoyancy
Chain	Poor	Average	High	NA[a]	Negative
Wire	Poor to good	Average	High	Slight	Negative
Nylon	Excellent	High	Low	High	Slightly negative
Polyester	Excellent	High	Moderate	Moderate	Negative
Polypropylene	Excellent	High	Low	Moderate	Positive

[a]Not applicable.

Multicomponent Slack Moorings

The mooring system suggestions that follow are made simply to give the wave energy conversion designer a "starting point." The reader is urged to consult the cited references for more in-depth discussions of deep water mooring systems. Referring to Figure 6.11, the mooring may have several segments. Ansari (1979) analyzes a three-segment mooring system, whereas Childers (1974a, 1974b, 1975) presents a more general discussion of deep water moorings. For deep ocean wave energy conversion terminals, a two-segment single-point slack mooring system consisting of a heavy chain for segment 1 (for strength and weight) and nylon line for segment 2 (for both flexibility and strength) is recommended. Since, from Table 6.1, both segments are negatively buoyant, catenary profiles of each segment will result. To minimize horizontal excursions, a clump weight should be attached where segments 1 and 2 are joined together.

Tension (Taut) Mooring

Referring to Figure 6.12, a three-segment mooring system is suggested. Segment 1 consists of a chain attached to the anchor for strength. From the results in Berteaux (1976), a polyester line would be excellent for segment 2 because of its excellent endurance to cyclic loading. The length of segment 2 will depend on the design "stretch" and the design spring constant k_c. Segment 3 should be either chain or wire. This segment must be strong and

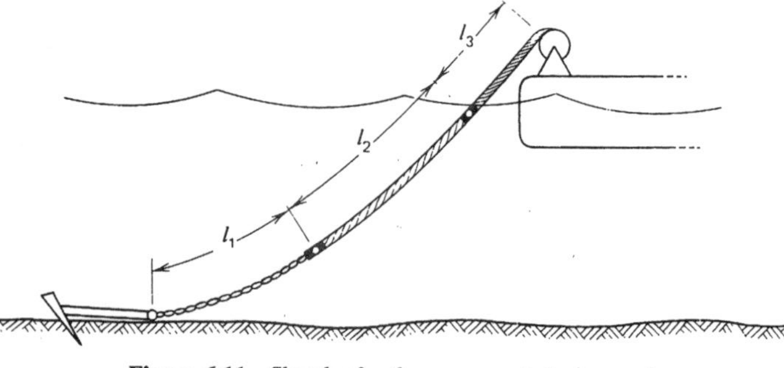

Figure 6.11 Sketch of a three-segment slack mooring.

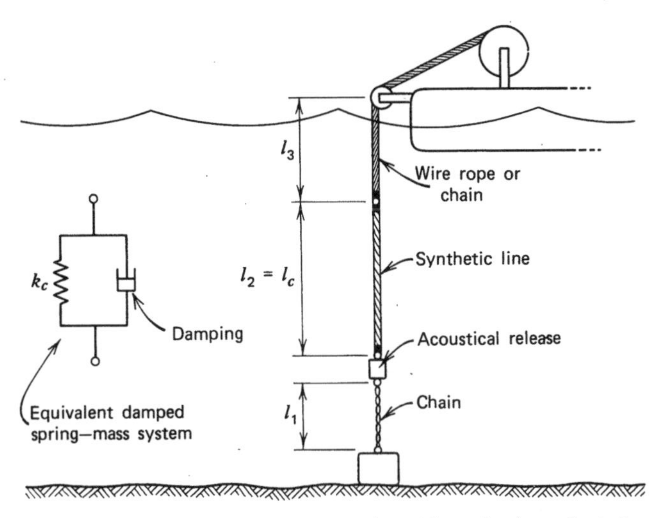

Figure 6.12 Sketch of a three-segment taut mooring with an elastic synthetic line segment.

should not snap under abrupt loading. The chain mass (if chain is chosen for segment 3) also affects f_z and f_θ since this mass increases both m and I_y in equations (6.8) and (6.9), respectively.

Example 6.3

The terminal sketched in Figure 6.9 when freely floating displaces 8×10^6 lb (3.56×10^7 N) of water where the depth is 1000 ft (305 m) and the wind speed is 20 knots (33.8 ft/sec or 10.3 m/sec). The diameter of the terminal is 100 ft (30.5 m) whereas the draft is 15.8 ft (4.82 m). The displaced mass is 2.48×10^5 lb-sec^2/ft (3.62×10^6 kg), and the added mass from Figure 4.3b is

$$
\begin{aligned}
m_w &= 0.167\rho D^3 \\
&= 0.167(2.00)100^3 \\
&= 3.34 \times 10^5 \text{ lb-sec}^2/\text{ft } (4.88 \times 10^6 \text{ kg})
\end{aligned}
$$

The natural heaving frequency for this freely floating body, from equation (4.3), is

$$
\begin{aligned}
f_z = \frac{1}{T_z} &= \frac{1}{2\pi}\sqrt{\frac{\rho g A_{wp}}{m + m_w}} \\
&= \frac{1}{2\pi}\sqrt{\frac{\rho g \pi D^2/4}{m + m_w}} \\
&= \frac{1}{2\pi}\sqrt{\frac{2.00(32.2)\pi(100)^2/4}{(2.48 + 3.34) \times 10^5}} \\
&= 0.148 \text{ Hz}
\end{aligned}
$$

or

$$T_z = 6.74 \text{ sec}$$

From the results in Figure 2.8, the reader can see that this heaving period is in the neighborhood of the peak wave energy spectrum for the 20-knot wind speed, which occurs at a period of 6.60 sec from the results of Example 2.9.

Assuming the freeboard to be equal to the draft, the natural pitching frequency in equation (4.6) is

$$f_\theta = \frac{1}{T_\theta} = \frac{1}{2\pi}\sqrt{\frac{C}{I_y + I_w}}$$

$$= \frac{1}{2\pi}\sqrt{\frac{\rho g \pi D^4/64}{\frac{m}{4}\left(\frac{D^2}{4} + 4d^2\right) + \frac{\rho \pi D^2 d^3}{12}}}$$

$$= \frac{1}{2\pi}\sqrt{\frac{gD^3}{dD^2 + 64d^3/3}}$$

$$= \frac{1}{2\pi}\sqrt{\frac{32.2(100)^3}{15.8(100)^2 + 64(15.8)^3/3}}$$

$$= 0.184 \text{ Hz} \tag{6.12}$$

or

$$T_\theta = 5.45 \text{ sec}$$

where the expression for I_y for the circular cylindrical hull having equal freeboard and draft is (Eshbach, 1975)

$$I_y = \frac{m}{4}\left(\frac{D^2}{4} + 4d^2\right) \tag{6.13}$$

From the results in Figure 2.8, where the wave period spectrum for a 20-knot wind is presented, the value of T_θ is somewhat away from the peak of the spectrum that occurs at $T_{(max)} = 6.60$ sec. Thus the pitching motions present less of a problem than the heaving motions.

We now determine the effects on the natural heaving and pitching frequencies of a four-point tension mooring. The anchors for these moorings should be either clump-type or explosive-type-embedment anchors. There are four lines extending from the hull to each anchor. These lines are 21-strand wire ropes of 1-in. (0.0254-m) diameter each, with a design modulus of elasticity Y_c of 2.9×10^7 lb/in^2 (2.0×10^{11} N/m^2). The unstretched length of each cable is 980 ft (299 m). From the results in equation (6.11), the value of the spring constant is

$$k_c = \frac{Y_c \pi r_c^2}{l_c}$$

$$= \frac{2.9 \times 10^7 \pi (0.5)^2}{980}$$

$$= 2.32 \times 10^4 \text{ lb/ft } (3.39 \times 10^5 \text{ N/m})$$

With this value, equations (6.8) and (6.9), respectively, yield the following:

$$f_z = \frac{1}{T_z} = \frac{1}{2\pi} \sqrt{\frac{\rho g \pi D^2/4 + nk_c}{m + m_w}}$$

$$= \frac{1}{2\pi} \sqrt{\frac{2.00(32.2)\pi(100)^2/4 + 16(2.32 \times 10^4)}{(2.48 + 3.34) \times 10^5}}$$

$$= 0.195 \text{ Hz}$$

or

$$T_z = 5.12 \text{ sec}$$

and

$$f_\theta = \frac{1}{T_\theta} = \frac{1}{2\pi} \sqrt{\frac{\rho g \pi D^4/64 + Dk_c(n/2)}{(\rho \pi D^2 d/16)(D^2/4 + 4d^2) + \rho \pi D^2 d^3/12}}$$

$$= \frac{1}{2\pi} \sqrt{\frac{2.00(32.2)\pi(100)^4/64 + 100(2.32 \times 10^4)8}{\dfrac{2.00\pi(100)^2(15.8)}{16} \Big/ \left[100^2/4 + 4(15.8)^2\right] + \dfrac{2.00\pi(100)^2 15.8^3}{12}}}$$

$$= 0.189 \text{ Hz}$$

or

$$T_\theta = 5.30 \text{ sec}$$

The effect of the 16 steel lines is to significantly raise the heaving frequency, while only slightly raising the pitching frequency.

In Example 6.3, after the embedment anchors are in place, tension is applied to the mooring lines by the winch, thereby increasing the displacement of the terminal and slightly altering the motion characterisitics of the floating body. For the purposes of this book, however, these alterations can be neglected.

B Coastal Zone Operation

In the relatively shallow water of the coastal zone both the satellite wave energy convertors and the terminal can be moored by using embedment anchors. The *mooring configuration* will normally be *slack* since the *tidal range* in the coastal zone may be large enough to prohibit the use of a tension mooring system. Furthermore, wave-induced motions of floating structures are much larger in shallow water than in deep water. For this reason a *multipoint* mooring system would be more apropos in shallow water, as illustrated in Figure 6.13. Depending on the water depth and the sea bed slope and composition, the wave energy conversion designer might also choose a *rigid structure.* This configuration would have the greatest stability in a sea but would probably cost much more than the slack mooring system.

The analysis of a slack mooring is rather complicated, even for the simplest configuration—the one-segment catenary. The analysis of the catenary is presented by Korbut and Herbert (1970), and numerical results from this analysis are in the paper by Gault and Cox (1974). Ansari (1979) presents an iterative scheme for solving the multisegment slack mooring problem. Referring to Figure 6.14 for notation, the relationship between the *line tensions* at the *float* F_0 and at the *anchor* F_a for a catenary configuration is

$$F_a = F_0 - w_c h \tag{6.14}$$

where w_c is the submerged weight of the mooring line per unit length. The

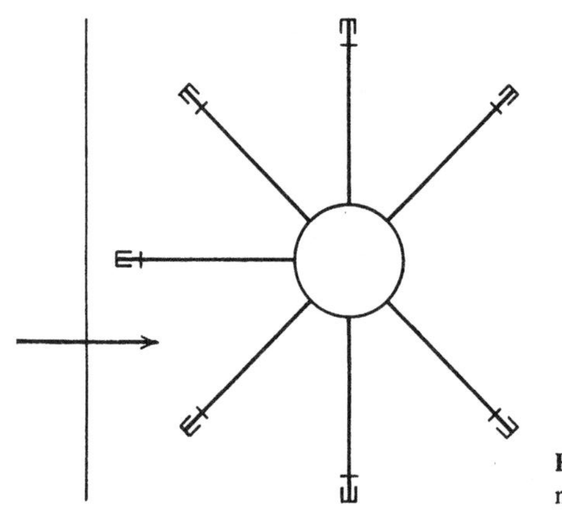

Figure 6.13 Plan view sketch of a slack mooring.

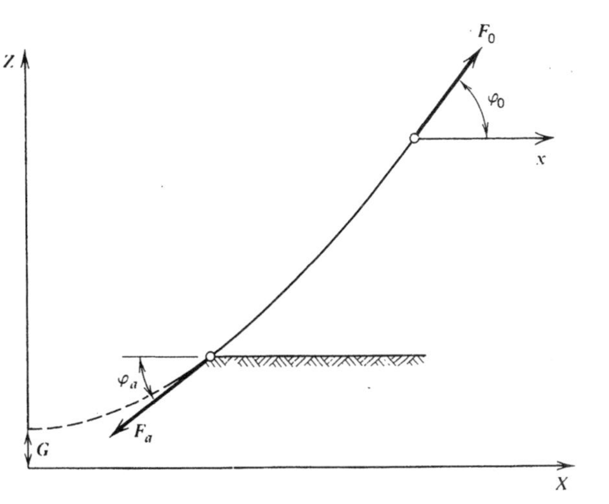

Figure 6.14 Notation for a catenary mooring configuration.

angles at the float φ_0 and the anchor φ_a are related by

$$\varphi_a = \cos^{-1}\left[\frac{\cos(\varphi_0)}{1 - w_c h/F_0}\right] \tag{6.15}$$

The bottom of the extended catenary is located above the origin of the X–Z coordinate system at a distance

$$G = \frac{F_0 \cos(\varphi_0)}{w_c} \tag{6.16}$$

The horizontal positions of the float and anchor are, respectively,

$$X_0 = G\cosh^{-1}\left[\frac{1}{\cos(\varphi_0)}\right] \tag{6.17}$$

and

$$X_a = G\cosh^{-1}\left[\frac{1}{G}\left(\frac{F_0}{w_c} - h\right)\right] \tag{6.18}$$

Finally, the length of the mooring line is

$$l_c = G\left[\sinh\left(\frac{x_0}{G}\right) - \sinh\left(\frac{X_a}{G}\right)\right] \tag{6.19}$$

The reader can see that the use of a computer is needed to determine the profile of the *slack* mooring.

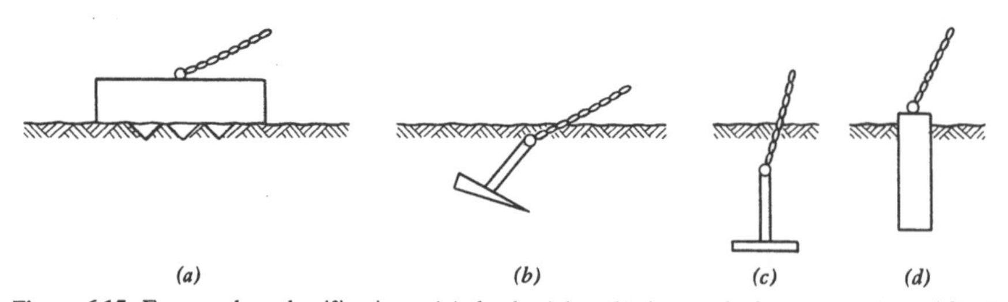

Figure 6.15 Four anchor classifications: (*a*) deadweight; (*b*) drag embedment; (*c*) plate; (*d*) pile.

The selection of the *anchor* depends on the water depth, the bed material, the bed slope, and the cost. There are four anchor groups, as discussed by Valent and Atturio (1977): the *deadweight*, the *drag embedment*, the *plate*, and the *pile* (these are sketched in Figure 6.15). For many slack mooring and taut mooring applications in deep water, the deadweight is used. The most commonly used anchor in intermediate and shallow is the drag embedment. This type of anchor is discussed by Berteaux (1976) in detail. An example of an embedment anchor is shown in Plate 6.2, which is the *Stato anchor* used in the Kaimei study. This is a rather massive anchor weighing several tons. Other embedment anchors include the *Navy Stockless*, the *Danforth*, the *Boss*, and the *Stimson*. The use of a slack mooring with a drag embedment anchor is illustrated in Example 6.4.

Example 6.4

As discussed by Berteaux (1977) the measure of the ability of a specific anchor to resist *breakout* in a certain type of soil is called the *holding*

Plate 6.2 One of the 20-ton anchors used on the Kaimei in the Sea of Japan. Courtesy of the Japan Marine Science and Technology Center.

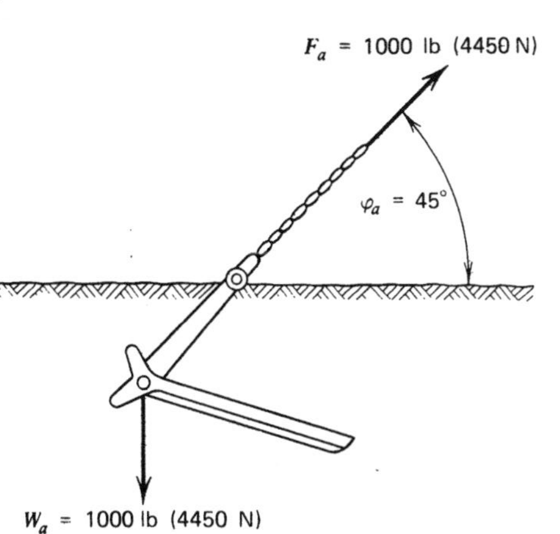

Figure 6.16 A sketch of the Stato anchor in example 6.4.

power K_a. This is defined as the ratio of the horizontal breakout force (experimentally determined) to the anchor weight W_a. In addition to the holding power of the anchor, the chain attached to the anchor will resist breakout if it is partially embedded. This aspect is discussed by Gault and Cox (1974).

Consider the situation sketched in Figure 6.16, where a 1000-lb (4450-N) Stato anchor is embedded in a soil for which the holding power is 15. The tension in the chain at the anchor F_a is 1000 lb (4450 N), and the chain angle φ_a is 45° to the horizontal (refer to Figure 6.14 for notation). The horizontal component of the tension F_a is

$$F_a \cos(\varphi_a) = 1000(0.707)$$

$$= 707 \text{ lb } (3140 \text{ N})$$

Since the holding power is 15, the maximum allowable horizontal force at this angle is

$$\left[F_a \cos(\varphi_a)\right]_{\max} = K_a W_a$$

$$= 15(1000)$$

$$= 15{,}000 \text{ lb } (66{,}700 \text{ N})$$

The vertical component of the tension

$$F_a \sin(\varphi_a) = 1000(0.707)$$

$$= 707 \text{ lb } (3140 \text{ N})$$

is rather less than the weight of the anchor W_a. The anchor weight and the vertical resistance of the soil above the anchor together give the anchor a vertical stability. Thus the anchor is stable under this loading condition.

The *steel chain* in this example is 1-in., with a linear weight density w_c of 10 lb/ft (146 N/m). The chain is attached to a wave energy conversion device located in 100 ft (30.5 m) of water. By knowing the conditions at the anchor, we can now determine the average horizontal force on the wave energy convertor that is assumed due to wind and waves. The tension at the convertor F_0 is obtained from equation (6.14); thus

$$\begin{aligned} F_0 &= F_a + w_c h \\ &= 1000 + 10(100) \\ &= 2000 \text{ lb } (8900 \text{ N}) \end{aligned} \tag{6.20}$$

where h is the water depth. The angle of the chain at the convertor is obtained from equation (6.15):

$$\begin{aligned} \varphi_0 &= \cos^{-1}\left[\frac{1 - w_c h}{F_0}\cos(\varphi_a)\right] \\ &= \cos^{-1}\left\{\left[1 - \frac{10(100)}{2{,}000}\right]\cos(45°)\right\} \\ &= 69.3° \end{aligned} \tag{6.21}$$

The average horizontal force on the convertor is then

$$\begin{aligned} F_0\cos(\varphi_0) &= 1000\cos(69.3) \\ &= 1000(0.353) \\ &= 353 \text{ lb } (1570 \text{ lb }) \end{aligned}$$

Mooring costs may make the cost of delivered converted wave power prohibitive. Thus the designer must constantly be cognizant of the trade-offs between mooring capacity and mooring costs. The reader is urged to consult the report edited by Dawson (1979) on this cost-effective aspect.

References

Anonymous (1977), "State-of-the-Art of Mooring and Dynamic Positioning," Technical Report, Frederick R. Harris, Inc., August.

Anonymous (1979), "Environmental Development Plan for Ocean Waves and Currents Energy Conversion," Technical Report, Energy and Environmental Analysis, Inc.

Ansari, K. A. (1979), "How to Design a Multi-component Mooring System," *Ocean Industry*, March, pp. 60–68.

Berteaux, H. O. (1976), *Buoy Engineering*, Wiley-Interscience, New York.

Bretschneider, C. L. (1967), "Fundamentals of Ocean Engineering"—Parts 1 through 8, *Ocean Industry*, June (1967) to June (1968).

Carmichael, A. D. (1978), "An Experimental Study and Engineering Evaluation of the Salter Cam Wave Energy Converter," Massachusetts Institute of Technology, Cambridge, Mass., Report No. MITSG 72-22, December.

Childers, M. A. (1974a), "Deep Water Mooring—Part I, Environmental Factors Control Station Keeping Methods," *Petroleum Engineer*, September, pp. 36–58.

Childers, M. A. (1974b), "Deep Water Mooring—Part II, Ultradeep Water Spread Mooring System," *Petroleum Engineer*, October, pp. 108–118.

Childers, M. A. (1975), "Deep Water Mooring—Part III, Equipment for Handling the Ultradeep Water Spread Mooring System," *Petroleum Engineer*, May, pp. 114–132.

Dawson, J. K. (1979), "Wave Energy," *Energy Paper 42*, Department of Energy (U.K.) AERE Harwell.

Dugger, G. L., Ed. (1979), "Ocean Thermal Energy for the 80's," *Proceedings, 6th OTEC Conference*, U.S. Department of Energy, Washington, D.C., June.

Eshback, O. W. (1975), *Handbook of Engineering Fundamentals*, Wiley, New York.

Gault, J. A., and Cox, W. R. (1974), "Method for Predicting Geometry and Load Distribution in an Anchor Chain from a Single Point Mooring Buoy to a Buried Anchor," *Proceedings, 6th Offshore Technology Conference*, Houston, Vol. II, May, Paper No. OTC 2062.

Heller, S. R. (1970), "The Cost Effectiveness of Natural and Synthetic Fiber Ropes in the Marine Environment," ASME Paper 70-WA/Unt-9.

Hong, S. T. (1974), "Tension in a Taut Line Mooring; Frequency Domain Analysis," *Proceedings, 6th Offshore Technology Conference*, Houston, Vol. II, May, Paper No. OTC 2069.

Kinsman, B. (1965), *Wind Waves*, Prentice-Hall, Englewood Cliffs, N.J.

Korbut, M. D., and Herbert, E. J. (1970), "Some Notes on Static Anchor Chain Curve," *Proceedings, 2nd Offshore Technology Conference*, Houston, Vol. I, May, pp. 147–160.

Lewis, L. F., and Waite, K. G. (1980), "Ocean Energy Technology: An Environmental Overview," preprint, *26th Meeting of the Institute of Environmental Science*, Franklin Institute, Philadelphia, Pa., May 12.

McCormick, M. E. (1973), *Ocean Engineerings Wave Mechanics*, Wiley-Interscience, New York.

McCormick, M. and Masuda, Y. (1980), "Review of the Wave Energy Conversion Project in the Sea of Japan," *Proceedings, 7th Ocean Energy Conference*, U.S. Department of Energy, Washington, D.C., June.

Miyazaki, T., and Masuda, Y. (1980), "Tests on the Wave Power Generator. 'Kaimei'," *Proceedings, 12th Offshore Technology Conference*, Houston, May, Paper No. OTC 3689.

Sorensen, R. M. (1978), *Basic Coastal Engineering*, Wiley-Interscience, New York.

Streeter, V. L. (1971), *Fluid Dynamics*, McGraw-Hill, New York.

U.S. Army (1973), *Shore Protection Manual*, Fort Belvoir, Va.

Valent, P. J., and Atturio, J. M. (1977), "Anchor Systems" *Proceedings, 4th OTEC Conference*, U.S. Department of Energy, New Orleans, La., March, pp. V-49 to V-55.

Appendix A*

Bibliography

A.1 Symposia Proceedings

The following are referred to under other headings by A.1a, A.1b, A.1c, and A.1d, respectively:

a. (1976), "Wave and Salinity Gradient Energy Conversion," Workshop, University of Delaware, Newark, Delaware, May.

b. (1978), "Wave and Tidal Energy," Symposium, British Hydromechanics Research Association, Canterbury, England, September.

c. (1979), "Symposium on Wave Energy Utilization," Chalmers University of Technology, Gothenburg, Sweden, October.

d. (1980), "7th Ocean Energy Conference," U.S. Department of Energy, Washington, D.C.,June.

A.2 Wave Mechanics

Kinsman, B. (1965), *Wind Waves—Their Generation and Propagation on the Ocean Surface*, Prentice-Hall, Englewood Cliffs, N. J.

Lighthill, J. (1978), *Waves in Fluids*, Cambridge University Press, Cambridge, England.

Stoker, J. J. (1957), *Water Waves*, Wiley-Interscience, New York.

A.3 Ocean Engineering Wave Mechanics

Bascom, W. (1964), *Waves and Beaches*, Doubleday, Garden City, N. Y.

Berteaux, H. O. (1976), *Buoy Engineering*, Wiley-Interscience, New York.

Brebbia, C. A., and Walker, S. (1979), *Dynamic Analysis of Offshore Structures*, Newnes-Butterworths, London.

Harikawa, K. (1978), *Coastal Engineering*, Wiley-Halsted, New York.

Ippen, A. T., Ed. (1966), *Estuary and Coastline Hydrodynamics*, McGraw-Hill, New York.

McCormick, M. E. (1973), *Ocean Engineering Wave Mechanics*, Wiley-Interscience, New York.

Muga, B. J., and Wilson, J. F. (1969), *Dynamic Analysis of Ocean Structures*, Plenum, New York.

*The following appendixes were written with the help of Messrs. Eric Midboe and Thomas Hudon of Gibbs and Cox, Inc., and Ensign Marc Rolfes of the U.S. Navy.

Muir-Wood, A. M. (1968), *Coastal Hydraulics*, Gordon and Breach, New York.

Newman, J. N. (1977), *Marine Hydrodynamics*, MIT Press, Cambridge, Mass.

Silvester, R. (1974), *Coastal Engineering*, Elsevier, New York.

Sorensen, R. M. (1978), *Basic Coastal Engineering*, Wiley-Interscience, New York.

U.S. Army (1973), *Shore Protection Manual*, U.S. Government Printing Office, Washington, D.C.

Wiegel, R. L. (1964), *Oceanographical Engineering*, Prentice-Hall, Englewood Cliffs, N. J.

A.4 General Wave Energy Conversion

Baird, W. F. (1968), "On Means of Utilizing the Energy of Wind-Waves," M. S. Thesis, Civil Engineering, Queen's College, Kingston, Ontario.

Bergdahl, L., Claeson, L., Falkemo, C., Farsherg, J., and Rylander, A. (1979), "The Swedish Wave Energy Research Program," Preprint, A.lc.

Brown, C. E., and Hsu, C. C. (1979), "A Survey of Ocean Wave Power Absorption Techniques," Tech. Report 8005-1, Hydronautics, Inc., Laurel, Maryland, November.

Count, B. M. (1979), "Exploiting Wave Power," *Spectrum* (IEEE), September, pp. 42–49.

Dawson, J. K. (1979), "Wave Energy," Energy Paper No. 42, U.K. Department of Energy, February.

Grove-Palmer, C. O. J. (1979), "The U.K. Wave Energy Program," Preprint, A.lc.

Grove-Palmer, C. O. J. (1980), "Development of Wave Energy in the United Kingdom," Paper II A/5, A.ld.

Hudon, T., Midboe, E., and Watts, J. (1980), "A Systematic Evaluation of Wave Energy Systems," Paper II A/3, A.ld.

Isaacs, J. D. (1979), "Ideas and Some Developments of Wave-Power Conversion, Dynamic Wave Absorption and Deep-Sea Mooring," Preprint, A.lc.

Isaacs, J., Castel, D., and Wick G. (1976), "Utilization of the Energy from Ocean Waves," *Ocean Engineering*, Vol. 3, No. 4, pp. 175–187.

Isaacs, J., Wick, G., and Schmitt, W. (1976), "Utilization of Energy from Ocean Waves," Paper F-1, A.la.

Katory, M., and Lacey, A. (1979), "Application of the Two-Dimensional Green's Function to the Hydrodynamic Analysis of Wave-Energy Devices," Preprint, A.lc.

Lachmann, B. (1979), "EUROCEAN Marine Energy Programme," Preprint. A.lc.

Longett-Higgins, M. S. (1976), "The Average Wave Forces Acting on Wave Power Machines," *Journal of the Society for Underwater Technology*, Vol. 2, No. 3, pp. 4–8.

Martin, M. D. (1974), "Power from Ocean Waves," American Society of Mechanical Engineers, Paper 74-WA/Pwr-5.

Mattisson, I. (1979), "A Wave Measuring Project for Wave Energy Prospecting Purpose," Preprint, A.lc.

McCormick, M. E. (1978), "Wind-Wave Power Available to a Wave Energy Converter Array," *Ocean Engineering*, Vol. 5, No. 2, pp. 67–74.

McCormick, M. E. (1978), "Wave Energy Conversion in a Random Sea," *Proceedings*, 13*th Intersociety Energy Conversion Engineering Conference*, San Diego, August.

McCormick, M. E. (1979), "Waves, Salinity Gradients and Ocean Currents—Alternative Energy

Sources," *Proceedings, 6th OTEC Conference*, U.S. Department of Energy, Washington, D.C., Paper 4, June.

McCormick, M. E. (1979), "Ocean Wave Energy Concepts," *Proceedings*, "*Oceans '79 Conference* (MTS–IEEE), San Diego, pp. 553–558.

Nath, J., and Williams, R. (1976), "Preliminary Feasibility Study of Utilization of Water Wave Energy," Paper I, A.la.

Panicker, N. N. (1976), "Power Resources Potential of Ocean Surface Waves," Paper J., A.la.

Panicker, N. N. (1976), "Review of the Technology for Wave Power Conversion," *Marine Technology Society Journal*, Vol. 10, No. 3, pp. 7–15.

Panicker, N. N. (1977), "Energy from Ocean Surface Waves," *Ocean Energy Resources*, Vol. 4, pp. 43–69.

Pleass, C. M. (1978), "The Use of Wave Powered Systems for Desalination: A New Opportunity," Paper D.1, A.lb.

Rae, J. (1977), "Theoretical Aspects of Wave Power," Report TP721, AERE, Harwell, Didcot (U.K.), October.

Rogalski, W., Midboe, E., Sherwood, W., and Szeto, F. (1979), "The State of the Art in Alternate Ocean Energy Systems," Society of Naval Architects and Marine Engineers, Chesapeake Division, Arlington, Virginia, December.

Ross, David (1979), *Energy from the Waves*, Pergamon Press, Oxford, England.

Shelpuk, B. (1980), "Ocean Wave Energy—Program Overview," Paper II A/1, A.ld.

Slotta, L. S. (1976), "Recoverable Wave Power Concepts," Paper H, A.la.

A.5 Wave Energy Resource

Baird, W., and Mogridge, G. (1976), "Estimates of the Power of Wind-Generated Water Waves at Some Canadian Coastal Locations," Report LTR-HY-53, Canadian Hydraulics Laboratory, Ottawa, August.

Baird, W. F. (1978), "Estimation of Wave Energy Using a Wing-Wave Hindcast Technique," Paper F3, A.lb.

Creech, Clayton (1977), "Five-Year Climatology (1972–1976) off Yaquina Bay, Oregon," Sea Grant Report, Oregon State University, Corvallis, Oregon, December.

Dutkiewicz, R., and Nurick, G. (1978), "Wave Energy off the Coast of South Africa," Paper Fl, A.lb.

Gran, Sverre (1977), "Estimates of Wave Power at the Coast of Norway," Report No. 77-570, Det. norske Veritas, Hovik, Norway, November.

Hogben, Neil (1979), "Methods for Estimating Power Available from Sea Waves," Final Report, International Energy Agency Wave Data Workshop, Oxford, U.K., September.

Hudspeth, R. T. (1976), "The Effect of Nonlinearities on Wave Power Estimates," Paper G, A.la.

Lazanoff, S. M. (1978), "Preliminary Verification of the FNWC Twenty Years Northern Hemisphere Wave Spectral Climatology," Report No. 7891, Hoffman Maritime Consultants, Glen Head, New York, December.

Pierson, W., and Salfi, R. (1976), "The Temporal and Spatial Variability of Power from Ocean Waves Along the West Coast of North America," Paper E, A.la.

Pierson, W., and Salfi, R. (1980), "A Northern Hemisphere Analysis of the Wave Energy Resource," Paper II A/2, A.ld.

Rylander, A. (1976), "Sea Waves at the Swedish Coastal Waters as a Source of Energy," Report No. SH 74-76, Chalmers University of Technology, Gothenburg, Sweden, September.

Svensson, J. (1979), "Wave Data from the Baltic and Its Computation out of Atmospheric Pressure Fields," Preprint, A.lc.

Swaan, W. A. (1976), "North Sea Waves as a Source of Energy for the Netherlands," Report No. 740-1, Bureau voor Scheepsbouw, Bloemendaal, Holland, June.

A.6 Heaving and Pitching Devices

Baz, A., and Morcos, S. (1978), "Development and Testing of a Direct Wave Energy Converter," Paper B.7, A.lb.

Blandino, A., Brighenti, A., and Vielmo, P. (1979), "Tecnomare's Contribution to Sea Wave Energy Exploitation," Preprint, A.lc.

Guenther, D., Jones, D., and Brown, D. (1979), "An Investigative Study of a Wave Energy Device," *Energy*, Vol. 4, pp. 299–306.

Jones, D., Guenther, D., and Chiou, W. (1980), "Power Extraction from Deep Ocean Waves Employing a Novel Wave Energy Device," American Society of Mechanical Engineers, Paper No. 80-Pet-29, February.

Mei, C., and Newman J. (1979), "Wave Power Extraction by Floating Bodies," Preprint, A.lc.

Slotta, L. S. (1980), "Power Extraction from a Transient Heaving Cylinder," Final Report, Slotta Engineering Associates, Corvallis, Oregon.

Srokosz, M., and Evans, D. (1978), "A Theory of Wave Power Absorption by Two Independently Oscillating Bodies," Paper B.1, A.lb.

A.7 Hydraulic and Pneumatic Devices

Bott, A., Hauley, J., and Hunter, P. (1978), "The Mauritius Sea-Wave Energy Project," Paper F.2, A.lb.

Grant, A. D. (1978), "Development of a Wave-Powered Marine Distress Beacon," Paper B.3, A.lb.

Hiramoto, A. (1978), "The Theoretical Analysis of an Air Turbine Generation System," Paper B.5, A.lb.

Hirsh, R. A. (1976), "Analog Simulation of a Wave Activated Turbine Generator Buoy," Report EW-9-75, U.S. Naval Academy, Annapolis, Maryland, January.

James, W. (1979), "Extraction of Power from Waves Using Harbour Resonators," Preprint, A.lc.

Lighthill, J. (1978), "Two-Dimensional Analyses Related to Wave-Energy Extraction by Submerged Resonant Ducts," *Journal of Fluid Mechanics*, Vol. 91, Part 2, pp. 253–317.

Masuda, Y. (1979), "Experimental Full Scale Results of Wave Power Machine Kaimei in 1978," Preprint, A.lc.

Masuda, Y., and Miyazaki, T. (1978), "Wave Power Electric Generation Study in Japan," Paper B.6, A.lb.

McCormick, M. E. (1974), "A Parametric Study of a Wave-Energy Conversion Buoy," *Offshore Technology Conference Proceedings*, Paper OTC 2125, May.

McCormick, M., Holt, R., and Bosworth, C. (1975), "A Pneumatic Wave-Energy Converter for Offshore Structures," *Offshore Technology Conference Proceedings*, Paper OTC 2259, May.

McCormick, M. E. (1974), "Analysis of a Wave-Energy Conversion Buoy," *Journal of Hydronautics* (AIAA), Vol. 8, No. 3, pp. 77–82.

McCormick, M., Carson, B., and Rau, D. (1975), "An Experimental Study of a Wave-Energy Conversion Buoy," *Marine Technology Society Journal*, Vol. 9, No. 3, pp. 39–42.

Miyazaki, T., and Masuda, Y. (1980), "Tests on the Wave Power Generator Kaimei," *Offshore Technology Conference Proceedings*, Paper OTC 3689, May.

Whittaker, T., and Wells, A. (1978), "Experiences with a Hydropneumatic Wave Power Device," Paper B.4, A.lb.

Wick, G., and Castel, D. (1978), "The Isaacs Wave-Energy Pump: Field Tests off the Coast of Kaneohe Bay, Hawaii," *Ocean Engineering*, Vol. 5, pp. 235–242.

A.8 Pressure Devices

French, M. J. (1979), "The Search for Low Cost Wave Energy and the Flexible Bag Devices," Preprint, A.lc.

Kayser, H. (1979), "A Submerged Wave Power Generator," Preprint, A.lc.

A.9 Particle Devices

Altmann, H., and Farley, F. (1979), "Latest Developments with the Tryslate Wave Energy Converter," Preprint, A.lc.

Farley, F., Parks, P., and Altmann (1978), "A Wave Power Machine Using Free Floating Vertical Plates," Paper B.2, A.lb.

Tornkvist, R. (1979), "Computer-Controlled Power from Irregular Ocean Waves," Preprint, A.lc.

A.10 Nodding Ducks

Carmichael, A. D. (1978), "An Experimental Study and Engineering Evaluation of the Salter Cam Wave Energy Converter," Report No. MITSG 78-22, Massachusetts Institute of Technology, Cambridge, Mass., December.

Mynett, A., Serman, D., and Mei, C. (1979), "Characteristics of Salter's Cam for Extracting Energy from Ocean Waves," *Applied Ocean Research*, Vol. 1, No. 1, pp. 13–20.

Salter, S. H. (1974), "Wave Power," *Nature*, Vol. 249, No. 5459, pp. 720–724.

Salter, S., Jeffrey, D., and Taylor, J. (1976), "The Architecture of Nodding Duck Wave Power Generators," *The Naval Architect*, January, pp. 21–24.

Serman, D., and Mei, C. (1980), "Note on Salter's Energy Absorber in Random Waves," *Ocean Engineering*, Vol. 7, No. 4.

A.11 Contouring Rafts

Haren, P. (1978), "Optimal Design of Hagen-Cockerell Raft," Master's Thesis, Massachusetts Institute of Technology, Cambridge, Mass. October.

Haren, P., and Mei, C. (1980), "Rafts for Absorbing Wave Power," 13th Naval Hydrodynamics Symposium, Preprint, Tokyo, October.

A.12 Wave Focusing

Ambli, N., Budal, K., Falnes, J., and Sørenssen, A. (1977), "Wave Power Conversion," Preprint, 10th World Energy Conference, Istanbul, September.

Anonymous (1979), "DAM–ATOLL—Ocean Wave Energy Extraction," Report LR-28932, Lockheed-California Company, Burbank, California, January 19.

Budal, K (1977), "Theory for Absorption of Wave Power by a System of Interacting Bodies," *Journal of Ship Research*, Vol. 21, No. 4, pp. 248–253.

Budal, K. and Falnes, J. (1975), "A Resonant Point Absorber of Ocean-Wave Power," *Nature*, Vol. 256, August 7, pp. 278–279.

Budal, K., Falnes, J., Kyllingstad, A., and Oltedal, G. (1979), "Experiments with Point Absorbers in Regular Waves," Preprint, A.lc.

Count, B., Collings, N., Davis, J., and Knott, G. (1978), "An Experimental Investigation of Point Absorbers," Report, Energy Technology Support Unit, AERE Harwell, U.K.

Duncan, J., and Brown, C. (1980), "Development of a Numerical Method for the Calculation of Power Absorption by Arrays of Similar Arbitrarily Shaped Bodies in a Seaway," Technical Report 8005-2, Hydronautics, Inc., Laurel, Maryland, March.

Evans, D. V. (1976), "A Theory for Wave-Power Absorption by Oscillating Bodies," *Journal of Fluid Mechanics*, Vol. 77, pp. 1–25.

Evans, D. V. (1979), "Some Theoretical Aspects of Three-Dimensional Wave-Energy Absorbers," Preprint, A.lc.

Falnes, J., and Budal, K. (1978), "Wave-Power Conversion by Point Absorbers," *Norwegian Maritime Research*, Vol. 6, No. 4, pp. 2–11.

Falnes, J. (1979), "Radiation Impedence Matrix and Optimum Power Absorption for Interacting Oscillators in Surface Waves," Preprint, A.lc.

Mehlum, E., and Stamnes, J. (1979), "On the Focusing of Ocean Swells and Its Significance in Power Production," Preprint, A.lc.

Newman, J. N. (1979), "Absorption of Wave Energy by Elongated Bodies," *Applied Ocean Research*, Vol. 1, No. 4, pp. 189–196.

Sebastiani, G., Berta, M., and Blandino, A. (1978), "Energy from Sea Waves: System Optimization by Diffraction Theory," *Proceedings*, *Oceans* '78 (MTS–IEEE), Washington, D.C., Paper 26C, September.

A.13 Energy Conversion

Binns, K. J. (1979), "The Use of Permanent Magnet Machines for Low Speed Generation," Preprint, A.lc.

Bishop, H., and Rees, G. (1979), "An Electrical Generation and Transmission Scheme for Wave Power," Preprint, A.lc.

Hidden, A. E. (1978), "The Kymatic Bridge: A Null-Balance Method for Testing Models of Wave Power Devices," Paper D.3, A.lb.

Thorborg, K. (1979), "Frequency Converter for Ocean Wave Energy Utilization," Preprint, A.lc.

Vimukta, D., Baker, T., and Plumpton, B. (1978), "Integrating Wave Power into the Electricity Supply System," Paper H.3, A.lb.

Wilson, M. N. (1979), "Slow Speed Generators with Superconducting Windings," Preprint, A.lc.

Appendix B

Some Wave Energy Conversion Patents

Some Wave Energy Conversion Patents

The purpose of this Appendix is to acquaint the reader with some of the patented devices in the categories discussed in Chapter 4. The reader will note that in most of the recent patents other patents are referenced on the cover page. For those interested in obtaining information on those patents, simply write to the U.S. Patent Office, 2021 Jefferson Davis Highway, Arlington, Virginia 22202.

Excerpts of the following patents are presented:

Heaving Devices

M. W. Gustafson and K. Loqvist, No. 3,965,364, 22 June 1976

A. H. Jackson, No. 4,091,618, 30 May 1978

Y. Solell, No. 4,145,885, 27 March 1979

M. Tornabene, No. 3,930,168, 30 December 1975

Pitching Devices

Y. Masuda, No. 3,204,110, 31 August 1965

Pressure Device

M. J. French, No. 4,164,383, 14 August 1979

Surging Device

S. R. Adams, No. 1,318,637, 14 October 1919

Particle Motion Device

R. E. Tornkvist, No. 4,036,563, 19 July 1977

Nodding Ducks

S. H. Salter, No. 3,928,967, 30 December 1975

E. Wood, No. 4,048,512, 13 September 1977

Wave Rafts

C. Cockerell, No. 4,098,084, 4 July 1978

G. E. Hagen, No. 4,077,213, 7 March 1978

Outrigger Device

K. Widecrantz and W. Gatton, No. 3,970,415, 20 July 1976

Pneumatic and Hydraulic Devices

Y. Masuda and I. Kanda, No. 3,200,255, 10 August 1965

H. A. Mattera, No. 3,870,893, 11 March 1975

G. W. Moody, No. 4,189,918, 26 February 1980

Focusing Device

I. Thorsheim, No. 4,172,689 30 October 1979

Energy Conversion

R. E. Salomon and S. M. Harding, No. 4,178,517, 11 December 1979

United States Patent [19]
Gustafson et al.

[11] **3,965,364**
[45] **June 22, 1976**

[54] WAVE GENERATION

[76] Inventors: **Manfred Wallace Gustafson**, Gamla Fagerstavagen 4; **Kaj-Ragnar Loqvist**, Regnbagsvagen 40, both of 773 00 Fagersta, Sweden

[22] Filed: **June 10, 1974**

[21] Appl. No.: **478,145**

[30] **Foreign Application Priority Data**

June 18, 1973 Sweden 7308523

[52] **U.S. Cl.** **290/53**;415/7; 417/331

[51] **Int. Cl.**^{2} **F03B 13/10**

[58] **Field of Search** 290/52, 53, 54, 42, 290/43; 415/7; 417/331, 332, 333

[56] **References Cited**

UNITED STATES PATENTS

1,898,973 2/1933 Lansing ...... 290/54 X
3,064,137 11/1962 Corbett et al... 290/54 X
3,126,830 3/1964 Dilliner........ 417/331
3,362,336 1/1968 Kadka 290/42
3,497,185 2/1970 Dively........... 415/7
3,640,514 2/1972 Albritton 415/7

Primary Examiner—Herman J. Hohauser
Attorney, Agent, or Firm—Larson, Taylor and Hinds

[57] **ABSTRACT**

A device for utilizing energy stored in wave motion. A buoyant body on the water surface is anchored so as to permit free, unrestricted vertical movement when acted upon by a heaving wave. An energy collecting member connected to the buoyant body and including propeller blades is located at a depth where the water is not subjected to the vertical wave motion.

7 Claims, 2 Drawing Figures

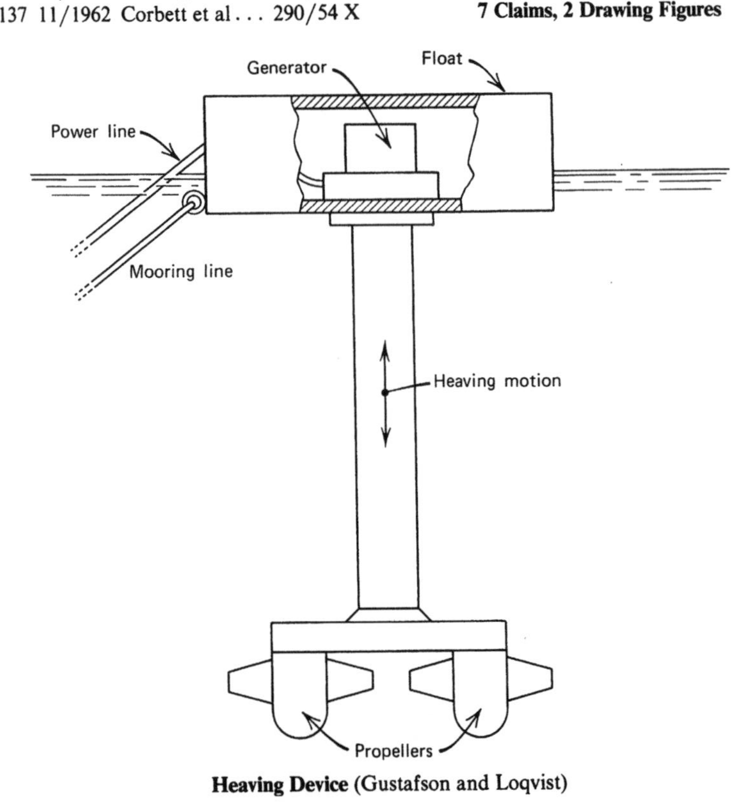

Heaving Device (Gustafson and Loqvist)

United States Patent [19]
Jackson

[11] **4,091,618**
[45] **May 30, 1978**

[54] **OCEAN MOTION POWER GENERATING SYSTEM**

[76] Inventor: **Arlyn H. Jackson,**
3128 Monterey St.,
Oxnard, Calif. 93030

[21] Appl. No.: **695,612**

[22] Filed: **June 14, 1976**

[51] **Int. Cl.**^{2} **E02B 9/03**
[52] **U.S. Cl.** **60/497;** 417/333
[58] **Field of Search** 60/325, 397, 398, 495–507; 417/330–333, 337; 185/4, 33

[56] **References Cited**

U.S. PATENT DOCUMENTS

1,560,425	11/1925	Huff	417/333
3,205,969	9/1965	Clark	60/398 X
3,307,827	3/1967	Silvers et al	60/501
3,487,228	12/1969	Kriegel	417/331
3,504,648	4/1970	Kriedt	60/398 X
3,664,125	3/1972	Strange	60/398 X
3,970,415	7/1976	Widecrantz et al	417/332

Primary Examiner—Allen M. Ostrager
Assistant Examiner—Stephen F. Husar
Attorney, Agent, or Firm—Warren T. Jessup

[57] **ABSTRACT**

A wave motion power generating system comprised of a floating buoy attached to a pump which pumps water from a container submerged below water level. The pump removes water from the container creating a hydrostatic head which can be used to draw water through a drive system for operating a power generator. The buoy may be in the form of a cylindrical ring mounted on the leg of an ocean platform with the pump being a cylindrical container also mounted on a leg well below the surface of the ocean. The motion created by ocean waves moves the buoy up and down operating the pump to create a void in a tank submerged well below the ocean surface, thus creating the hydrostatic head. This hydrostatic head can be used to drive a turbine which in turn operates a power generator.

6 Claims, 4 Drawing Figures

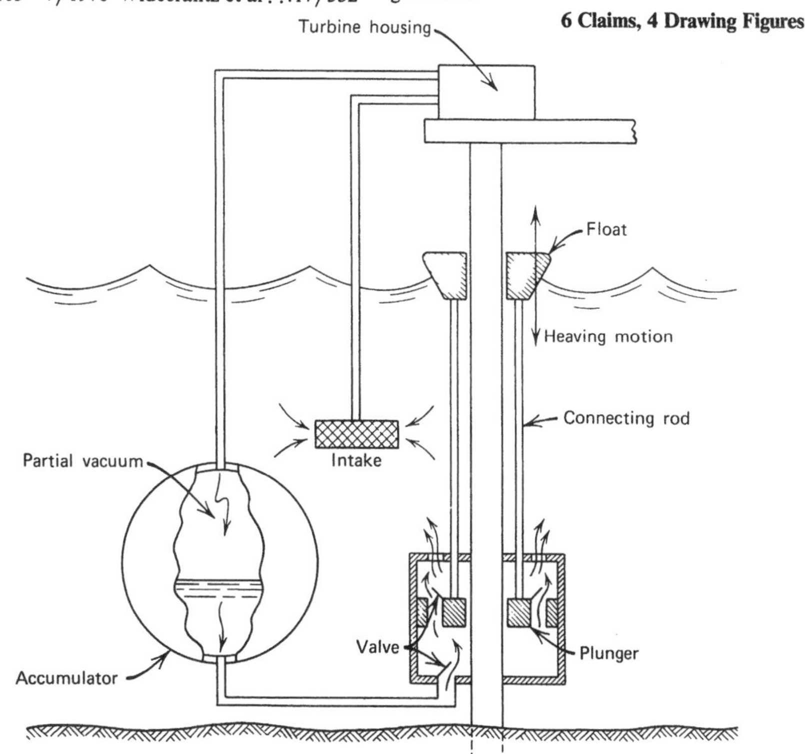

Heaving Device (Jackson)

United States Patent [19] **Solell**

[11] **4,145,885**
[45] **Mar. 27, 1979**

[54] **WAVE MOTOR**

[76] Inventor: **Yedidia Solell,** Rav-Ashi 1, Tel Aviv, Israel

[21] Appl. No.: **854,226**

[22] Filed: **Nov. 23, 1977**

[30] **Foreign Application Priority Data**

Sep. 23, 1977 [IL] Israel 52982
Oct. 21, 1977 [IL] Israel 53179

[51] **Int. Cl.**^{2} **F03B 13/12**
[52] **U.S. Cl.** **60/504;** 60505; 60/507; 60/398; 290/53
[58] **Field of Search** 60/398; 495, 497, 502, 60/504, 505, 507, 716; 61/20; 185/2, 33; 290/42, 53; 417/100, 330, 331, 337

[56] **References Cited**

U.S. PATENT DOCUMENTS

366,768 7/1887 Elias . . . . 61/20 X
415,812 11/1889 Dowe 60/504
971,343 9/1910 Barr 60/497
1,292,303 1/1919 Garwood 60/504

Primary Examiner—Edgar W. Geoghegan
Attorney, Agent, or Firm—Benjamin J. Barish

[57] **ABSTRACT**

A wave motor is described comprising a float, a displaceable member coupled to the float so as to be displaced by the ascent and descent of the float, a pair of shafts, and a transmission including a pair of one-way clutches coupling the displaceable member to the shafts to rotate one in one direction during the ascent of the float and to rotate the other in the opposite direction during the descent of the float. In one described embodiment, the displaceable member is a wheel which is partially rotated in opposite directions by the ascent and descent of the float; and in a second described embodiment, the displaceable member is a rack which is moved upwardly by the ascent of the float and downwardly by its descent.

7 Claims, 6 Drawing Figures

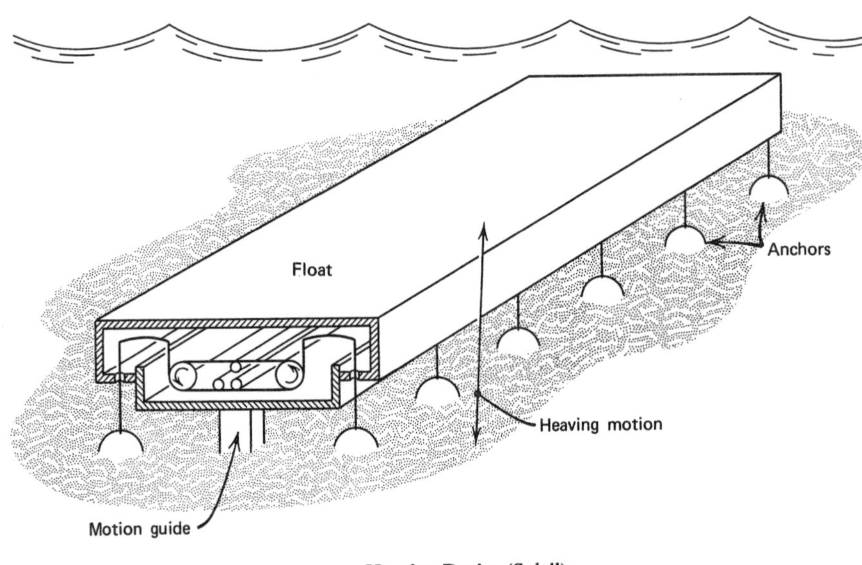

Heaving Device (Solell)

United States Patent [19]
Tornabene

[11] **3,930,168**
[45] **Dec. 30, 1975**

[54] **WAVE-ACTION POWER APPARATUS**

[76] Inventor: **Michael G. Tornabene,** 462 7th Ave., New York, N.Y. 10018

[22] Filed: **Apr. 1, 1974**

[21] Appl. No.: **457,075**

Related U.S. Application Data

[63] Continuation-in-part of Ser. No. 428,349, Dec. 26, 1973, and a continuation-in-part of Ser. No. 432,211, Jan. 10, 1974

[52] **U.S. Cl.** **290/53**; 417/331
[51] **Int. Cl.^{2}** **F03B 13/10**
[58] **Field of Search** 290/42, 43, 53, 54; 417/330, 331, 332, 333, 334

[56] **References Cited**

UNITED STATES PATENTS

953,600	3/1910	Edens	417/333
961,401	6/1910	Bonney	417/331
975,157	11/1910	Quedens	290/42
1,393,472	10/1921	Williams	290/42
1,396,580	11/1921	Kilcullen	417/333
1,864,499	6/1932	Grigsby	290/42
3,394,658	7/1968	Johnson	417/333
3,808,445	4/1974	Bailey	290/53

Primary Examiner—Robert K. Schaefer
Assistant Examiner—John W. Redman

[57] **ABSTRACT**

In a preferred embodiment of the invention, there is provided a double-action piston water

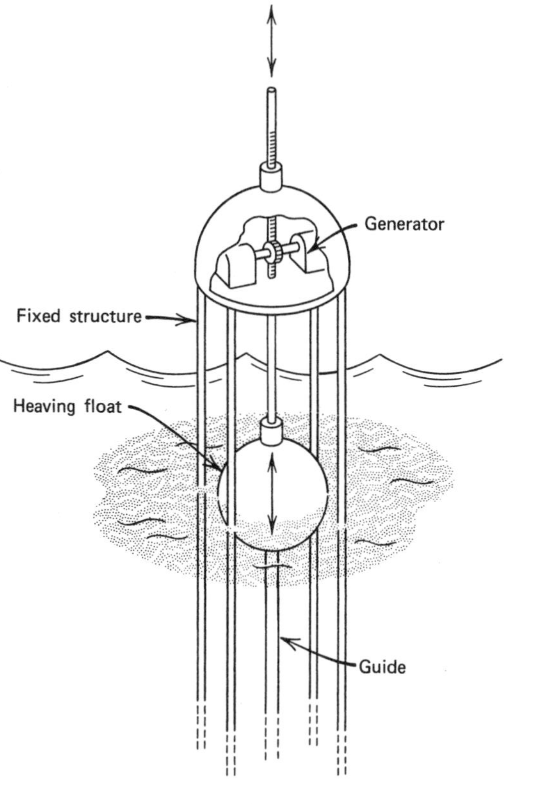

Heaving Device (Tornabene)

pump of elongated shape with the piston supported on a lever rod extending coaxially to the piston chamber and extending through both ends of the cylinder providing thereby equal volume displacement constantly throughout the cylinder inner space irrespective of the position of the piston during the to and fro strokes, with there being mounted on a lower end of the lever rod a float element revolvable around the lever rod with the lever rod as a central axis thereto and with there being detachably mounted on the float element an additional dense mass for varying the depth that the float element sits in the water and varying momentum and inertia during a stroke, the float element including separable and dismountable upper and lower halves and including bearing mountings of the upper and lower halves with a lubrication port and channel for pressurized lubrication of the bearing mountings, and there also being included in closed flow cycle inlet and outlet conduits to and from the opposite ends of the piston cylinder with appropriate one-way valves therein and mounted within the conduit cycle there being a turbine propelled by the pumped fluid such as pumped water, the cylinder and the conduits and the turbine being arranged relative to one-another to obtain a substantially unbroken circle of flow for accentuating fly-wheel-like inertia and momentum of the circularly flowing water, there also being an additional mass body mounted on a lower end of the lever rod having a stabilizing effect against distorting torques on the lever rod, the revolving float element also reducing any such distorting torque effects also, the rod lever being slidably supported by two bearing mountings located both above the upper extremity of the upward stroke of the lever rod responsive to the crest of a wave pushing upwardly on the float element.

22 Claims, 13 Drawing Figures

United States Patent Office

3,204,110
Patented Aug. 31, 1965

3,204,110
OCEAN WAVE ELECTRIC GENERATOR

Yoshio Masuda, 31–1 Kodanjutaku, 540 Ueda, Hinomachi, Minamitamagun Tokyo, Japan

Filed June 26, 1962
Ser. No. 205,427
Claims priority, application Japan July 7, 1961, 36/23,718;
Mar. 20, 1962, 37/10,404
7 Claims.
(Cl. 290—42)

This invention relates to the ocean wave electric generator which changes the force of ocean waves into electric power.

The object of this invention is to supply electric power to buoys on the ocean which will be available for oceanographic study, meteorolotical observation, fairway buoy, marine product industry, fishing and military patrol, etc.

Storage batteries have been used for buoys until recently, but as their lives are relatively short, such buoys can not work for a long time without restoring the batteries. By using this ocean wave electric generator in which the electric power is generated by the wave energy, the buoys can be operated semi-permanently as they automatically charge batteries.

There are two different types of the ocean wave electric generator and their constructions are different according to the each application. Type **1** is suitable for small power plant of 1–10 watts and type **2** is suitable for a little larger power plant of 50–500 watts.

In order that the invention may be understood more clearly and carried into effect readily, embodiments thereof will now be described in detail, by way of example, with reference to the accompanying drawing in which:

FIGURE 1 as a cross sectional side view showing the construction of an ocean wave of electric generator type **1**,

FIGURE 2 is an explanatory view showing the motion of the ocean wave electric generator of FIGURE 1,

FIGURE 3 is a cross sectional side view showing the construction of an ocean wave electric generator type **2**,

FIGURE 4 is an explanatory view showing the motion of the ocean wave electric generator of FIGURE 3,

FIGURE 5 is a partly sectional view of a gear mechanism which is used for this ocean wave electric generator,

FIGURE 6 is a cross sectional view of coaster wheels which are used for the above gear mechanism.

A buoy on the ocean is given very strong forces from the water motion of wave. One of the forces is buoyancy from the vertical motion of water surface, and the other is impacting force from the horizontal motion of water, but it is necessary to use an opposite force from the bottom of the sea in order to generate electric power from these forces. In a shallow sea, the opposite force is given by a mooring anchor, but it is very difficult to use the opposite force by the mooring anchor because of tide and very stormy weather. In a deep sea, it is impossible to get the opposite force by the mooring anchor.

This invention offers a solution to these problems. The ocean wave electric generators shown in FIGURES 1 and 3 generate electric power without the opposite force from the bottom of the sea.

The ocean wave electric generator type **1** is suitable for small unit and FIGURE 1 shows the cross sectional side view of its one example. It consists of buoy **1**, supporter **2**, center shaft **3**, arm **4**, pendulum weight **5**, electric generator **6**, small gear **7**, large gear **8**, wire **9**, submerging inertia body **10**, and backward weight **11**. The submerging inertia body **10** is connected by the wire **9** to one side of the buoy **1**, and the backward weight **11** is fixed to the other side of the buoy **1**. The weight of the submerging inertia body **10** in the water is the same as that of the backward weight **11**, but as the submerging inertia body **10** includes sea water in its body its inertia is very large. The buoy can keep its balance in still water. The center shaft **3** is fixed to the supporter **2** in the buoy **1**. The pendulum weight **5** is supported with the arm **4** to the center shaft **3**, and the arm **4** can rotate around the center shaft **3**. The electric generator **6** is installed in the pendulum weight **5**, and its shaft has the small gear **7**. The small gear **7** is engaged with the large gear **8** of the buoy **1**. Therefore, the rotation of the arm **4** around the center shaft **3** causes a high rotation of the shaft of the electric generator **6** as shown in FIGURE 2 through a gear mechanism which will be explained later.

The buoy **1** moves up and down by the force of buoyancy, but the submerging inertia body **10** is almost fixed by its large inertia, so the buoy **1** inclines by the wave motion. This phenomenon can be explained by the differences of each force to the buoy **1**, the submerging inertia body **10**, the backward weight **11**, and wave motion.

At first I must explain something about the wave motion. When a wave comes, the vertical position x, the vertical velocity dx/dt and the vertical acceleration of wave d^3x/dt^2 are shown in the following formulae 1–3.

$$x = H/2 \cdot \sin \omega t \tag{1}$$

$$dx/dt = \pi H/T \cdot \cos \omega t \tag{2}$$

$$d^2x/dt^2 = -2\pi^2 H/T^2 \cdot \sin \omega t \tag{3}$$

in which H is wave height and T is wave period.

The wave energy is transmitted by the force of viscosity between each particle of water, so its vertical force F_1 is shown the Formula 4.

$$F_1 = M \cdot d^2x/dt^2 = -M \cdot 2\pi^2 H/T^2 \cdot \sin \omega t \tag{4}$$

in which M is nearly equal to the mass of water particle.

On the other hand, the buoyancy force F_2 to the buoy by the wave is shown by the Formula 5.

$$F_2 = S \cdot x = S \cdot H/2 \cdot \sin \omega t \tag{5}$$

in which S is the horizontal sectional area of the buoy **1**. The phase of F_1 and F_2 is different at 180 degrees.

It is a very importatnt character, that the buoyancy force F_2 is given to the center of the buoy **1** and that this force is separated to each board of the buoy; its one part to the submerging inertia body **10** and the other part to the backward weight **11**. The mass of the submerging inertia body **10** is much larger than that of the backward weight **11**, so the submerging inertia body **10** is sunk into deep position by the wire **9**. The wave motion is limited to the shallow position of sea water, so the submerging inertia body **10** is not influenced by the force F_1 but it is influenced by the force F_2. As the result, the phase of its motion to wave surface is different about 180 degrees. On the other hand, the mass of the backward weight is small, therefore, the phase of its motion to wave surface comes near to zero degrees difference.

As the result of these phase differences, the buoy is inclined compulsorily and has no relation to free surface of wave. This is a very important phenomenon, because the force which inclines the buoy by free surface of wave is not strong enough to produce a large electric power in a small buoy without the submerging inertia body **10**. When the buoy **1** inclines as shown in FIGURE 2, the pendulum weight **5** does not incline, and relative rotation between the pendulum weight **5** and buoy **1** arises. The electric generator is rotated by this relative rotation by the principle mentioned above and it generates electric power. From the latest experiment of this type in which the buoy is about 200 liter, it is confirmed that this type can generate electric power of about 10 watts by a wave of the sea.

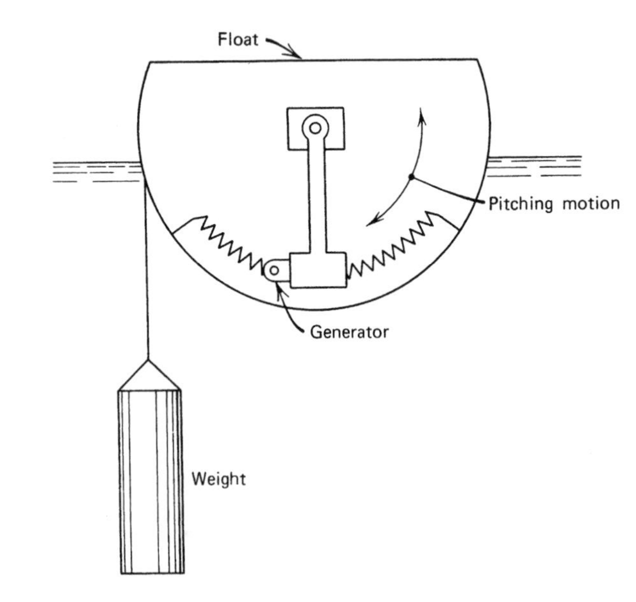

Pitching Device (Masuda)

United States Patent Office

3,200,255
Patented Aug. 10, 1965

3,200,255
OCEAN WAVE ELECTRIC GENERATOR

Yoshio Masuda, Tokyo, Japan
assignor of one-half to
Ichiro Kanda,
Kawaguchi, Saitamaken, Japan

Filed Jan. 24, 1961,
Ser. No. 84,555
Claims priority, application Japan,
Feb. 10, 1960, 35/4,274;
Oct. 14, 1960, 35/41,199
15 Claims. (Cl. 290—42)

This invention relates to a device which changes the force of ocean waves into electric power.

One of the objects of this invention is to supply electric power to buoys or offshore stations on the ocean which have the purpose of oceanographic study, meteorological observation, fairway buoy, marine product industry, fishing and military patrol, etc. Storage batteries have been used for buoys until recently, but as their lives are relatively short, such buoys cannot work for a long time without restoring the batteries. By using this "ocean wave electric generator" in which the electric power is generated by the wave energy, the buoys and offshore stations can be operated semipermanently as they automatically charge batteries.

Another object of this invention is to produce a large electric energy for island country where the ocean wave is one of the natural energy sources. It must be borne in mind that the wave energy exists at every point on the ocean and that it has much higher energy density than other natural energy sources such as wind, sunbeam or tide. At the coast of the Japan Sea, the wave height exceeds 1.2 meters for more than 240 days of the year, and the power of wave per 1 meter length of the seashore is about 30∼100 kw. in the stormy weather. A large ocean wave electric generator or a large wave motor is a kind of the mobile breakwater which breaks waves by utilizing its energy. This is another object of this invention.

There are three different types of the ocean wave electric generator and its construction is different according to the individual object and application.

Type **1** is suitable for deep sea, type **2** is suitable for shallow sea, and type **3** is suitable for small buoy. The advantages and other objects of this invention will be apparent from the following description and the attached drawing.

In the drawing,

FIGURE 1 is a cross sectional view of the ocean wave electric generator type **1**.

FIGURE 2 and FIGURE 3 show the mechanism of four vanes and air turbine of type **1**.

FIGURE 4 is a side view showing the mooring mechanism of buoy to the very deep sea.

FIGURE 5 is a cross sectional side view showing a construction of a mobile breakwater and its mooring mechanism.

FIGURE 6 is a top view of a mobile breakwater.

FIGURE 7 is a cross sectional side view showing a construction of an ocean wave electric generator type **2**.

FIGURE 8 is a top view showing an ocean wave electric generator of type **2**.

FIGURE 9 is a cross sectional side view showing a construction of an ocean wave electric generator type **3**.

FIGURE 10 is an explanatory view showing the motion of buoy with an ocean wave electric generator type **3**.

A buoy on the ocean is given very strong forces from the water motion of wave. One of the forces is buoyancy by the vertical motion of water surface, and the other is impacting force by the horizontal motion of the bottom of the sea in order to generate an electric power by these forces.

In a sallow sea, the opposite force is given by a mooring anchor, but it is very difficult to moor a large platform in very smooth weather. In a deep sea, it is impossible to get the opposite force by a mooring anchor.

This invention gives a solution to these problems. The ocean wave electric generator type **1** shown in FIGURE 1 generates electric power without the opposite force from the bottom of the sea. It consists of upper room **1**, connection room **2**, air pump room **3**, lower room **4**, weight **5**, air pipe **6**, vanes and turbine **7**, electric generator **8** and storage battery **9**. The disposition of each part is shown in FIGURE 1. The upper room **1** is made water-tight, and keeps air pipe **6** and storage battery **9** in it, and puts vanes and turbine **7** and electric generator **8** on it. The connection room **2** is situated between rooms **1** and **4**; its horizontal sectional area is small, and it is also made watertight. The air pump room **3** is attached under the upper room **1**, and its bottom opens to sea water through gape, but its head opens to the air through the air pipe **6** and the vane and turbine **7**. The lower room **4** is made watertight, and has a large buoyancy. The weight **5** is put under the lower room **4**. The buoy stands vertically by the weight of **5** and the buoyancy of the rooms **2** and **4**. The air pipe **6** connects the air pump room **3** with open air through the vanes and turbine **7**. The vanes and turbine **7** have four vanes and turbines which are shown in FIGURES 2 and 3. The electric generator **8** is connected to the turbine of **7**, and its electric output is supplied to load through the storage battery **9**. These operations will be explained later.

The buoyancy of the rooms **2** and **4** is a little larger than the total weight of the apparatus, so the sea level is kept about the middle of the connection room **2**. If this buoy is sunk into water and released, it will oscillate up and down, and the period T_2 of this oscillation is shown by the Formula 1.

$$T_2 = \frac{2\pi}{\sqrt{\frac{(S_1 + S_2)\rho}{M}}} \qquad (1)$$

T_2: the period of natural oscillation
π: 3.14
S_1: horizontal section area of the connection room **2**
S_2: horizontal section area of the air pump room **3**
ρ: weight of a unit volume of sea water
M: total mass which includes the additional water surrounded by this buoy

When ocean wave comes, the buoy is moved up and down by wave force. When the wave period T_1 is shorter than T_2, this buoy is almost kept immovable according to its large mass, and the relative vertical motion arises between this buoy and the wave surface. If T_1 is equal to T_2, the motion is resonant, and the relative vertical motion between this buoy and the wave surface gets bigger. If T_1 is no longer than T_2, the abovementioned relative vertical motion does not arise.

The air in the air pump room **3** is compressed and expanded by the water surface as the result of this relative vertical motion between this buoy and wave surface. Therefore, the air flows from the air pump room **3** to the open air and from the open air to the air pump room **3**. The vanes and turbine **7** operate by this air stream. FIGURES 2 and 3 show their actions.

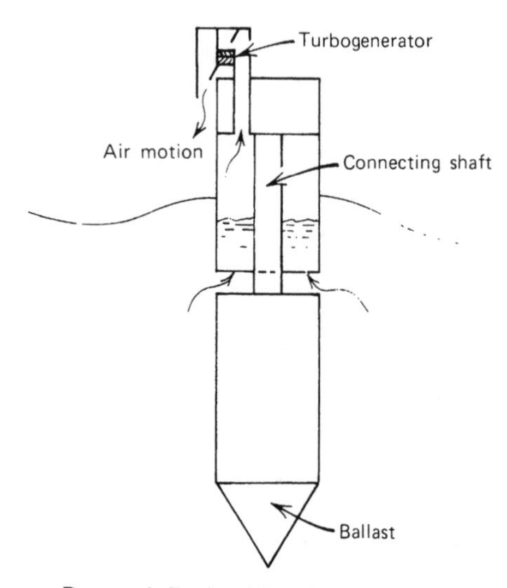

Pneumatic Device (Masuda and Kanda)

United States Patent [19] [11] **3,870,893**

Mattera [45] **Mar. 11, 1975**

[54] **WAVE OPERATED POWER PLANT**

[76] Inventor: **Henry A. Mattera,** 736 Fern St., Yeadon, Pa. 19050

[22] Filed: **Oct. 15, 1973**

[21] Appl. No.: **406,545**

[52] **U.S. Cl.** **290/53,** 290/42
[51] **Int. Cl.** **F03b 13/12**
[58] **Field of Search** ...... 290/53; 54, 43, 44; 417/330, 331, 332, 333, 334, 335, 336, 337.

[56] **References Cited**

UNITED STATES PATENTS

1,448,029 3/1923 Larry et al. 290/53
3,064,137 11/1962 Corbett et al. 290/53
3,200,255 8/1965 Masuda 290/53

Primary Examiner—G. R. Simmons
Attorney, Agent, or Firm—Zachary T. Wobensmith, 2nd; Zachary T. Wobensmith, III

[57] **ABSTRACT**

A wave operated power plant is described wherein a buoyant vessel is anchored on the ocean surface with the wave motion forcing ocean water up through vertical pipes in the hull causing rotation of blades in the pipes thereby through shafts rotating electrical generators and generating electricity, the water exhausted from the tubes being discharged onto a deck above the ends of the pipes and to the ocean.

4 Claims, 3 Drawing Figures

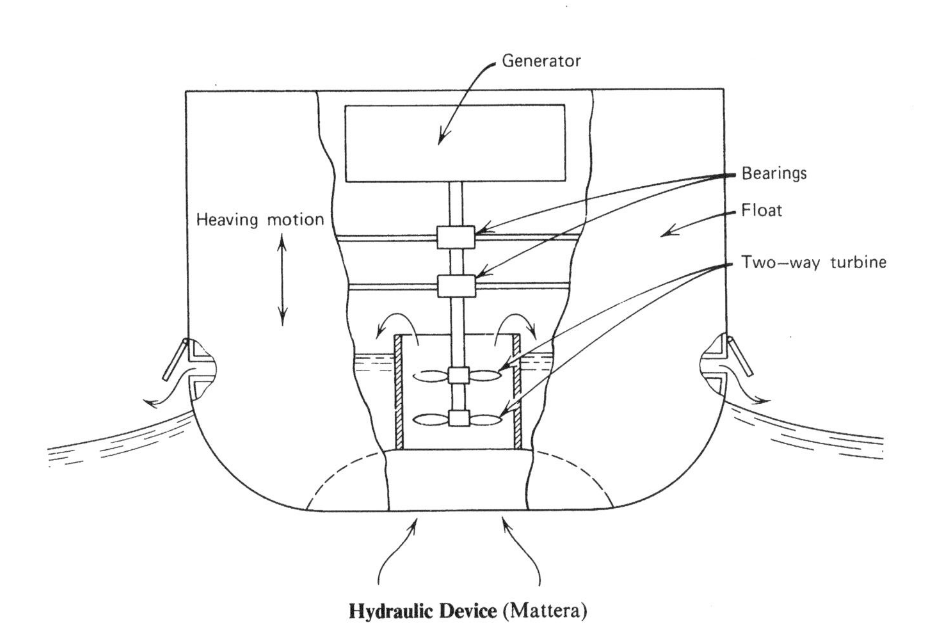

Hydraulic Device (Mattera)

United States Patent [19] **Moody et al.**

[11] **4,189,918**

[45] **Feb. 26, 1980**

[54] **DEVICES FOR EXTRACTING ENERGY FROM WAVE POWER**

[75] Inventors: **George W. Moody**, East Kilbride; **Robert A. Meir**, Cambustang, both of Scotland

[73] Assignee: **The Secretary of State for Energy in Her Britannic Majesty's Government of the United Kingdom of Great Britain and Northern Ireland,** London, England

[21] Appl. No.: **933,671**

[22] Filed: **Aug. 14, 1978**

[51] **Int. Cl.**^{2} **F03B 13/12**

[52] **U.S. Cl.** **60/398;** 290/53; 405/76; 417/330

[58] **Field of Search** 60/398; 495, 502; 290/42, 53; 405/76; 417/100, 330, 331, 332, 337

[56] **References Cited**

U.S. PATENT DOCUMENTS

568,117	9/1896	Price	60/398
3,928,967	12/1975	Salter	60/495 X
4,009,396	2/1977	Mattera et al	290/53
4,139,984	2/1979	Moody et al	60/398

OTHER PUBLICATIONS

Paper by D. V. Evans, Entitled "A Theory for Wave Power Absorption by Oscillating Bodies", 11th Symposium on Naval Hydrodynamics, vol 5, 1976, pp. 15–27.

Primary Examiner—Edgar W. Geoghegan
Attorney, Agent, or Firm—Larson, Taylor & Hinds

[57] **ABSTRACT**

The invention provides a device for extracting energy from waves on a liquid upon which the device is adapted to float. The device is allowed to move in response to the waves, and has a shape below the surface of the liquid, position of center of gravity, and value of radius of gyration about the center of gravity, in a vertical plane aligned in the direction of propagation of the waves, adapted so that the device in response to the waves inhibits to a substantial extent the transmission and/or reflection of waves by the device itself.

8 Claims, 12 Drawing Figures

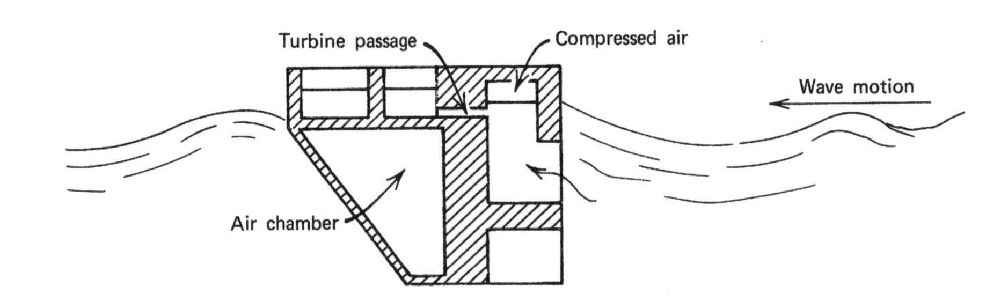

Hydraulic Device (Moody)

United States Patent [19] — [11] **4,164,383**

French — [45] **Aug. 14, 1979**

[54] **WATER WAVE ENERGY CONVERSION DEVICE USING FLEXIBLE MEMBRANES**

[76] Inventor: **Michael J. French, United Kingdom Atomic Energy Authority, 11 Charles II St., London, United Kingdom, S.W.1**

[21] Appl. No.: **799,524**

[22] Filed: **May 23, 1977**

[30] **Foreign Application Priority Data**
May 26, 1976 [GB] United Kingdom 21768/76

[51] **Int. Cl.**[2] **F04B 35/00**

[52] **U.S. Cl.** **417/330;** 60/398; 290/53

[58] **Field of Search** 60/398; 502, 504, 505; 61/20; 290/42, 53, 40; 417/100, 330; 185/33

[56] **References Cited**

U.S. PATENT DOCUMENTS

1,791,239 2/1931 Braselton 417/100
3,353,787 11/1967 Semo 60/398
3,961,863 6/1976 Hooper .... 290/42 X
3,989,951 11/1976 Lesster et al. .. 417/330 X

Primary Examiner—Edgar W. Geoghegan
Attorney, Agent, or Firm—Larson, Taylor and Hinds

[57] **ABSTRACT**

A device for conversion of sea wave energy comprises partially inflated bag-like enclosures formed from flexible impermeable material. The enclosures are connected via non return valves to a high pressure conduit and a low pressure return conduit so as to act as an air pump or bellows as the sea rises and falls around the enclosures. A turbine extracts energy from the high pressure conduit, the exhaust returning to the low pressure return conduit.

3 Claims, 13 Drawing Figures

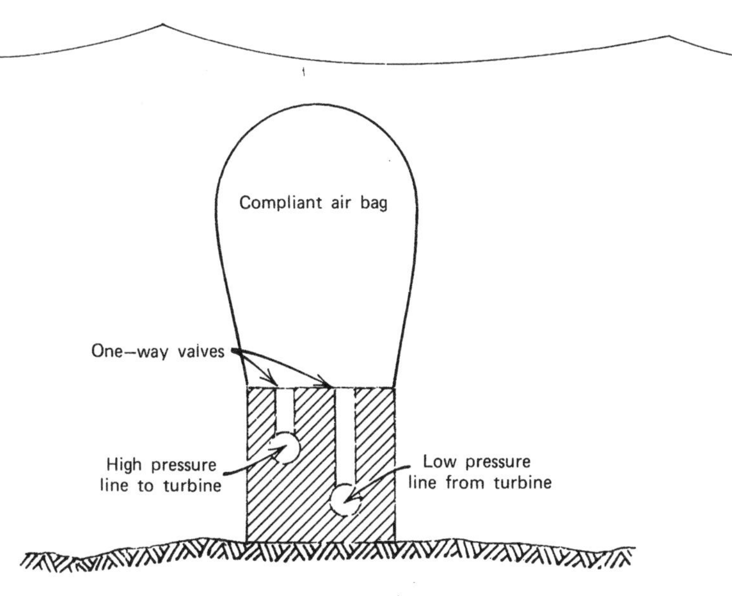

Pressure Device (French)

UNITED STATES PATENT OFFICE

SHERMAN R. ADAMS, OF LONG BEACH, CALIFORNIA.

WAVE-MOTOR

1,318,637 Specification of Letters Patent. **Patented Oct. 14, 1919.**

Application filed June 6, 1918. Serial No. 238,544.

To all whom it may concern:

Be it known that I, SHERMAN R. ADAMS, a citizen of the United States, residing at Long Beach, in the county of Los Angeles and State of California, have invented new and useful Improvements in Wave-Motors, of which the following is a specification.

My invention relates to wave motors and consists in the novel features herein shown, described and claimed.

Figure 1 is a side elevation of a pier equipped with a wave motor embodying the principles of my invention.

Fig. 2 is a vertical cross-section on the line 2—2 of Figure 1.

Fig. 3 is a horizontal sectional detail on the lines 3—3 of Figs. 1 and 2.

Fig. 4 is an enlarged fragmentary vertical longitudinal sectional detail on a plane parallel with Fig. 1 and on the lines 4—4 of Figs. 2 and 3.

Fig. 5 is a fragmentary cross-section on the line 5—5 of Fig. 1.

The pier platform 1 is mounted upon the piles 2, 3, 4, 5 and 6. The pier platform 1 extends a considerable distance straight out into the ocean from the shore. The piles 2 are channel-shaped in cross-section with the channels 7 and 8 facing each other. Bars 9 and 10 are slidingly mounted in the channels 7 and 8 and gear racks 11 and 12 are formed upon the inner faces of the bars 9 and 10. Bearings 13 and 14 are formed at the lower ends of the bars 9 and 10 and a stub-shaft 15 is loosely mounted in these bearings. Flanged drum wheels 16 and 17 are fixed upon the ends of the shaft 15 against the inner faces of the bearings 13 and 14 by set-screws 18 and 19. A bevel gear 20 is formed upon the outer face of the drum wheel 16.

The pier platform 1 connects the upper ends of the piles 2. A drive shaft 21 is mounted in a bearing 22 fixed upon the platform 1 and the shaft extends downwardly inside of the gear rack 11 slidingly through a bearing 23, said bearing 23 being fixed to the bar 9. A bevel opinion 24 is slidingly splined upon the shaft 21 in mesh with the bevel gear 20. The pinion 24 is held to go up and down with the bar 9 and gear rack 11 while the shaft 21 is held against endwise movement by the bearing 22 and is free to rotate. A countershaft 25 is mounted upon the platform 1 and is connected to the drive shaft 21 by bevel gears 26 and 27. The driven shaft 28 is mounted longitudinally of the platform 1 and connected to the countershaft 25 by gears 29 and 30.

The piles 3 are channel-shaped in cross-section similar to the piles 2, and bars 31 and 32 are slidingly mounted in the channels 33 and 34, and gear racks 35 and 36 are formed upon the inner faces of the bars 31 and 32. A turnbuckle brace 37 connects the lower ends of the bars 31 and 32. Trunnions 38 and 39 are fixed in the bars 31 and 32 above the brace 37 and drum wheels 40 are rotatably mounted upon these trunnions 38 and 39 against the inner faces of the bars 31 and 32. In a like manner a pair of drum wheels 42 is mounted upon bars 43 slidingly mounted in the piles 4, and drum wheels 44 are mounted upon bars 45 slidingly mounted in the piles 5, said drum wheels 40, 42 and 44 being in horizontal alinement with the upper sides of the drum wheels 16 and 17 and serving as guide pulleys. The drum wheels 46 are mounted upon sliding bars 46′ in the piles 6, said drum wheels 46 being larger than the drum wheels 16 and 17 and the tops of the wheels 46 being on a level with the tops of the wheels 16, 17, 40, 42 and 44, and the bottoms of the wheels 46 being considerably below the bottoms of the wheels 16 and 17. The cable 47 runs around the drum wheel 16 and one of the drum wheels 46 and over one of each pair of guide drum wheels 40, 42 and 44, and the cable 48 runs around the drum wheel 17 and one of the drum wheels 46 and the other ones of the guide drum wheels 40, 42 and 44.

Folding feathering impellers 49 connect the cables 47 and 48 crosswise at suitable distances apart, said impellers being mounted to be opened and pushed by the incoming waves and by the outgoing undertow. In other words, the impellers 49 face the shore on a line between the upper sides of the wheels 46 and the upper sides of the wheels 16 and 17 and face away from the shore on the return line from the lower faces of the wheels 16 and 17 to the lower faces of the wheels 46. As the cables 47 and 48 are driven by the impellers 49 the wheels 16 and 17 will be rotated to operate the bevel gear 20, drive the pinion 24, rotate the shaft 21, and drive the countershaft 25 to operate the driven shaft 28, and power may be taken from the driven shaft 28 in any suitable way and for any desired purpose.

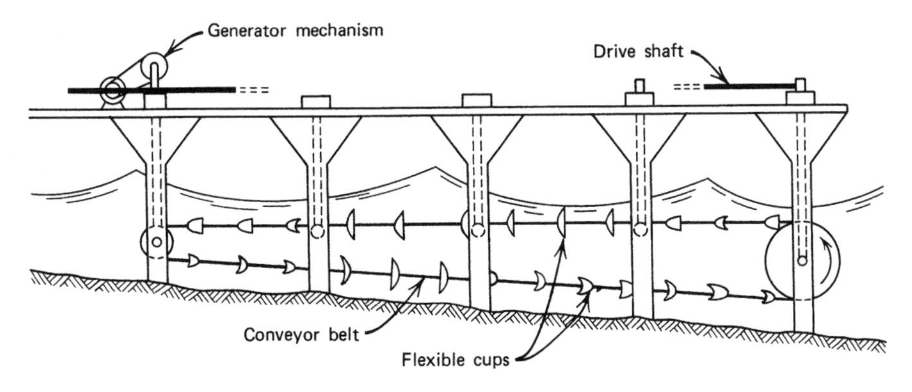

Surging Device (Adams)

United States Patent [19] | [11] **4,036,563**

Tornkvist | [45] **July 19, 1977**

---

[54] **WAVE MOTOR COMPRISED OF A SUBMERGED FLOATING NETWORK OF CHAMBERS FORMED BY WALLS PERMITTING VARIABLE GEOMETRY**

[76] Inventor: **Rolf E. A. Tornkvist,** Ritobergsvagen 8–16 L, 00300 Helsingfors 33, Finland

[21] Appl. No.: **696,821**

[22] Filed: **June 16, 1976**

Related U.S. Applications Data

[63] Continuation-in-part of Ser. No. 554,770, Jan. 28, 1975, abandoned.

[30] **Foreign Application Priority Data**

Feb. 5, 1974 Finland 312/74
Jan. 9, 1975 Finland 750052/75

[51] **Int. Cl.^{2}** **F04B 17/00;** F03B 13/12
[52] **U.S. Cl.** **417/331;** 60/398; 60/500; 290/53
[58] **Field of Search** **417/330, 331,** 332; 60/398, 499, 500, 504 505; 290/42, 53

[56] **References Cited**

U.S. PATENT DOCUMENTS

882,883 3/1908 Hillson 60/500
1,408,094 2/1922 Kersey 60/500
3,151,564 10/1964 Rosenberg 60/499
3,603,804 9/1971 Casey 417/332
3,758,788 9/1973 Richeson 417/332
3,961,863 6/1976 Hooper 60/499

Primary Examiner—Carlton R. Croyle
Assistant Examiner—Thomas I. Ross
Attorney, Agent, or Firm—Cushman, Darby & Cushman

[57] **ABSTRACT**

A wave energy transformer for transformation of wave energy into pressure energy of water in a pipe system. The transformer consists of a submerged space network of chambers at least partially closed by walls pivoted to adjacent walls. The walls are deformable due to the action of the waves. At least two of said deformable walls in each chamber are connected to at least one pumping means for pumping water into said pipe system owing to the deformation of said walls.

8 Claims, 12 Drawing Figures

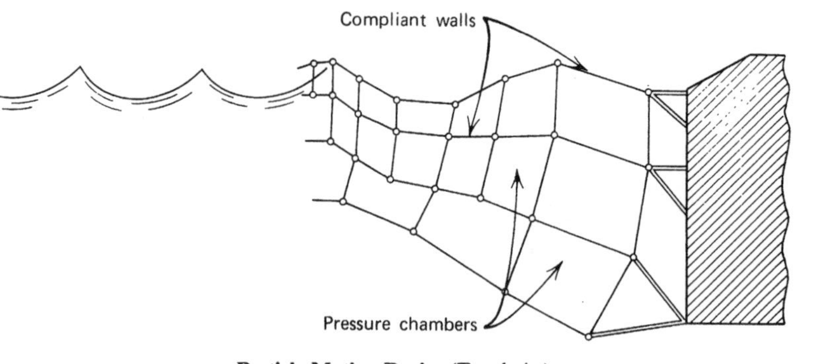

Particle Motion Device (Tornkvist)

United States Patent [19]
Salter

[11] **3,928,967**
[45] **Dec. 30, 1975**

[54] APPARATUS AND METHOD FOR EXTRACTING WAVE ENERGY

[76] Inventor: **Stephen Hugh Salter**, 143 E. Trinity Road, Edinburgh, EH53 PP, Scotland

[22] Filed: **Nov. 6, 1974**

[21] Appl. No.: **521,385**

[30] **Foreign Application Priority Data**
Nov. 15, 1973 United Kingdom 53119/73
May 6, 1974 United Kingdom 19763/74

[52] U.S. Cl. **60/398**; 60/495; 415/7; 417/332

[51] **Int. Cl.**[2] **F01D 25/00**; F03G 7/00; F04B 35/00

[58] **Field of Search** 60/495–507, 60/398; 415/2, 7; 417/330–334, 337; 290/42, 43, 52, 53

[56] **References Cited**

UNITED STATES PATENTS

1,016,022 1/1912 Lundquist........415/7
1,035,993 8/1912 Moore ...... 417/334 X
1,068,283 7/1913 Starry .... 417/334 X
1,074,292 9/1913 Reynolds.....417/334 X
1,263,865 4/1918 Dale .... 415/7 X

Primary Examiner—Allen M. Ostrager
Attorney, Agent, or Firm—Hill, Gross, Simpson, Van Santen, Steadman, Chiara & Simpson

[57] **ABSTRACT**

Apparatus and method for extracting the tremendous energy from a wave pattern as for example on the surface of a body of water which comprises a plurality of movable members so shaped that the surface of the member which engages the incoming wave causes the member to rotate about a substantially horizontal axis and remove energy from the wave and wherein the rear portion of the movable member is constructed so as to move with minimum energy transfer between the movable member and the fluid. The energy extracted by the movable member is converted into a hydraulic, electrical, mechanical, or chemical energy so as to allow the wave energy to be directly converted into useable form.

18 Claims, 6 Drawing Figures

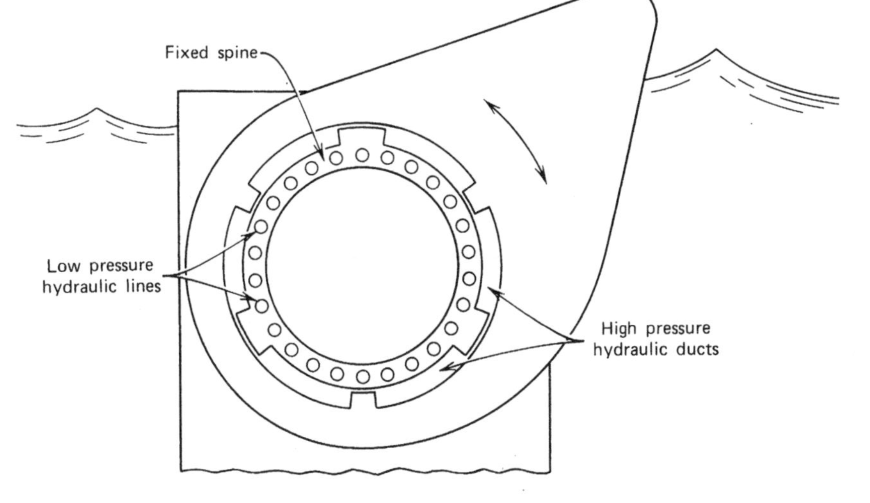

Nodding Duck (Salter)

United States Patent [19] [11] **4,048,512**

Wood [45] **Sept. 13, 1977**

[54] **SYSTEM FOR GENERATING POWER FROM WAVE MOTIONS OF THE SEA**

[75] Inventor: **Eric Wood**, Ossett, England

[73] Assignee: **Insituform (Pipes & Structures) Ltd.,**, England

[21] Appl. No.: **656,043**

[22] Filed: **Feb. 6, 1976**

[30] **Foreign Application Priority Data**

Feb. 7, 1975	United Kingdom	05254/75
Feb. 8, 1975	United Kingdom	05453/75
Apr. 23, 1975	United Kingdom	16740/75
May 1, 1975	United Kingdom	18131/75
May 1, 1975	United Kingdom	18132/75
May 14, 1975	United Kingdom	20247/75
June 12, 1975	United Kingdom	25287/75
Aug. 16, 1975	United Kingdom	34185/75

[51] **Int. Cl.**[2] **F03B 13/10**; F03B 13/12

[52] **U.S. Cl.** **290/53;** 290/42; 417/332; 417/337; 60/500; 60/501; 60/505

[58] **Field of Search** 290/42, 43, 53, 54; 417/231, 330, 331, 332, 337; 416/84–86; 60/495–497, 398, 501, 500, 505; 185/30, 33; 415/7

[56] **References Cited**

U.S. PATENT DOCUMENTS

3,928,967 12/1975 Salter.......... 417/332

OTHER PUBLICATIONS

Salter, "Wave Power," Nature Magazine, vol. 249, June 21, 1974, pp. 720–724.

Primary Examiner—Robert K. Schaefer
Assistant Examiner—Michael K. Mutter
Attorney, Agent, or Firm—Bierman & Bierman

[57] **ABSTRACT**

A power generating system for the generation of energy from the wave motions of the sea is buoyantly supported by the sea. The system comprises spine elements defining a spine on which ducks are rockably mounted, the system further including tension cable means pressing the spine elements towards each other thereby to define a spine, and energy conversion means for converting the rocking motion into electrical power.

11 Claims, 16 Drawing Figures

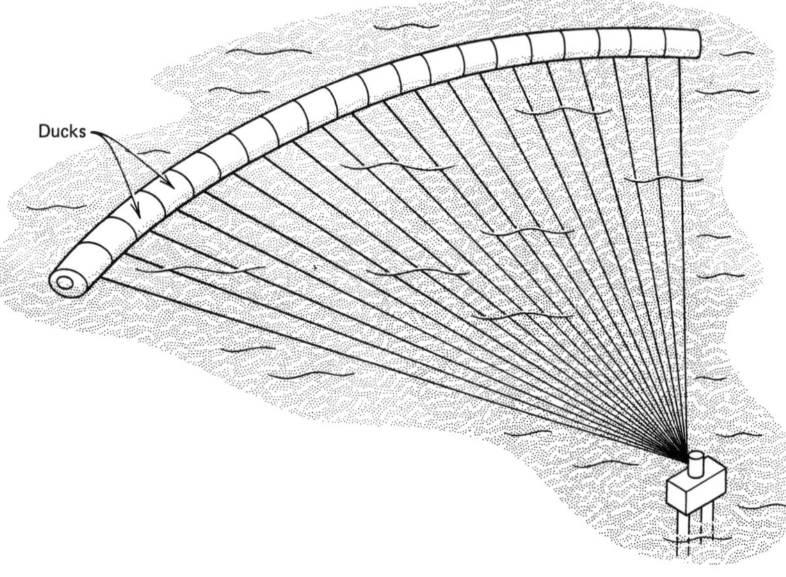

Nodding Duck (Wood)

United States Patent [19]

Cockerell

[11] **4,098,084**

[45] **Jul. 4, 1978**

[54] **APPARATUS FOR EXTRACTING ENERGY FROM WAVE MOVEMENT OF THE SEA**

[75] Inventor: **Christopher Cockerell,** Southampton, England

[73] Assignee: **Wavepower Limited,** Southampton, England

[21] Appl. No.: **678,863**

[22] Filed: **Apr. 21, 1976**

[30] **Foreign Application Priority Data**

Apr. 28, 1975 [GB] United Kingdom 17597/75

[51] **Int. Cl.**^{2} **E02B 9/08**

[52] **U.S. Cl.** **60/500**; 60/501; 417/332

[58] **Field of Search** 60/499–502, 60/505, 506; 417/330–333; 290/53

[56] **References Cited**

U.S. PATENT DOCUMENTS

879,992 2/1908 Wilson 417/331

901,117 10/1908 McManus .... 60/505 X

3,758,788 9/1973 Richeson ..... 60/500 X

Primary Examiner—Allen M. Ostrager

Assistant Examiner—Stephen F. Husar

Attorney, Agent, or Firm—McNenny, Pearne, Gordon, Gail, Dickinson & Schiller

[57] **ABSTRACT**

This invention relates to apparatus for extracting energy from movement of water, particularly sea waves, the apparatus comprising a plurality of buoyant members which are interconnected one with another so as to be movable relative to one another, each buoyant member being provided with a plate or plate-like member which is supported from the buoyant member and positioned so that in use of the apparatus it is submerged below the level of the water, and means are provided for converting the relative movement of the buoyant members into useful energy, such as electricity.

18 Claims, 31 Drawing Figures

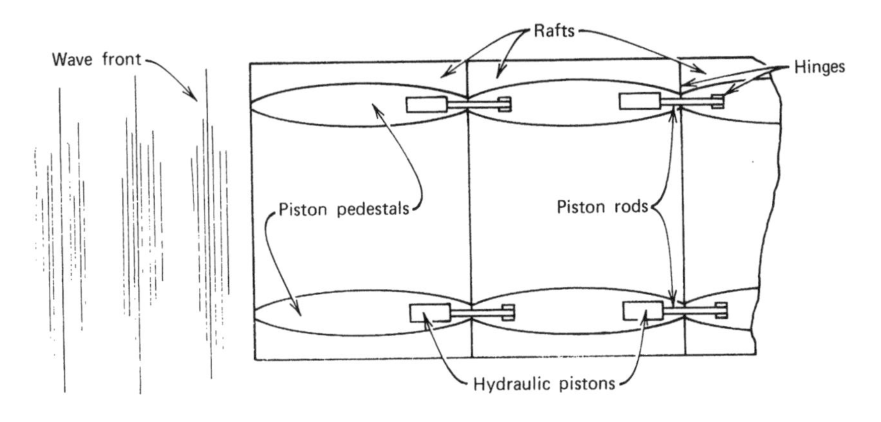

Wave Rafts (Cockerell)

United States Patent [19] [11] **4,077,213**

Hagen [45] **Mar. 7, 1978**

[54] **WAVE DRIVEN GENERATOR**

[75] Inventor: **Glenn E. Hagen**, New Orleans, La.

[73] Assignee: **Williams, Inc.**, New Orleans, La.

[21] Appl. No.: **657,892**

[22] Filed: **Feb. 13, 1976**

[51] **Int. Cl.²** **F03G 7/08**

[52] **U.S. Cl.** **60/500**; 60/501; 417/331

[58] **Field of Search** 60/497, 500, 501, 505, 60/506; 417/331, 332, 333, 337; 290/53

[56] **References Cited**

U.S. PATENT DOCUMENTS

882,883	3/1908	Hillson	60/501 X
917,411	4/1909	Casella et al.	60/501 X
1,567,470	12/1925	Settle	60/501 X
3,487,228	12/1969	Kriegel	417/331 X
3,758,788	9/1973	Richeson	60/500 X

Primary Examiner—Allen M. Ostrager
Attorney, Agent, or Firm—Arthur M. Dula; Ned L. Conley; Murray Robinson

[57] **ABSTRACT**

A plurality of different sized floats are connected into an array through nonlinear interfaces so their relative motions drive hydraulic pumping means. Floats in the array are sized to present a "black body" to the ocean waves incident upon the array. Fluid moved by the pumping means is used to drive an electric turbogenerator.

30 Claims, 9 Drawing Figures

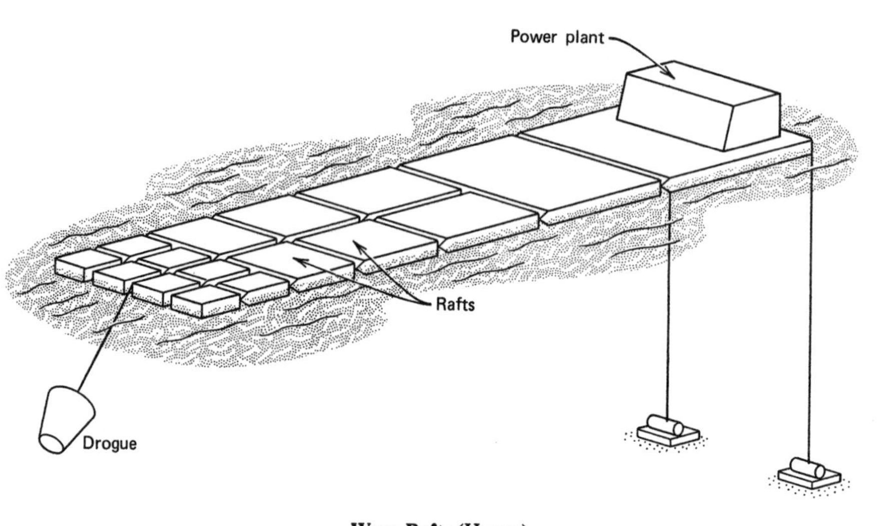

Wave Rafts (Hagen)

United States Patent [19] [11] **3,970,415**

Widecrantz et al. [45] **July 20, 1976**

[54] **ONE WAY VALVE PRESSURE PUMP TURBINE GENERATOR STATION**

[76] Inventors: **Kaj Widencrantz**, P.O. Box 72; **William R. Gatton**, P.O. Box 222, both of Port Republic, N.J. 08241

[22] Filed: **Apr. 10, 1975**

[21] Appl. No.: **566,794**

[52] **U.S. Cl.** **417/332**; 60/496; 290/42

[51] **Int. Cl.**^{2} **F04B 17/00**; F04B 35/00

[58] **Field of Search** 417/331, 332, 333; 60/496, 506; 290/42

[56] **References Cited**

UNITED STATES PATENTS

1,223,184 4/1917 Larson 417/332
2,935,024 5/1960 Kofahl ....60/496 X
3,126,830 3/1964 Dillner 417/331
3,603,804 9/1971 Casey ....417/332 X

FOREIGN PATENTS OR APPLICATIONS

562,217 8/1958 Canada 417/332

Primary Examiner—Carlton R. Croyle
Assistant Examiner—Richard E. Gluck
Attorney, Agent, or Firm—Richard L. Miller

[57] **ABSTRACT**

A new type of power generating plant that utilizes the motion of ocean waves to drive turbine generators in a power station; the plant including a series of underwater units each of which includes a hollow sphere that floats upon the water so that it rises and falls as waves move by, the bail being mounted on an end of a pivoting arm to which there is connected a piston slidable in a cylinder so to pump ocean water through a duct to the turbines in the power station.

2 Claims, 6 Drawing Figures

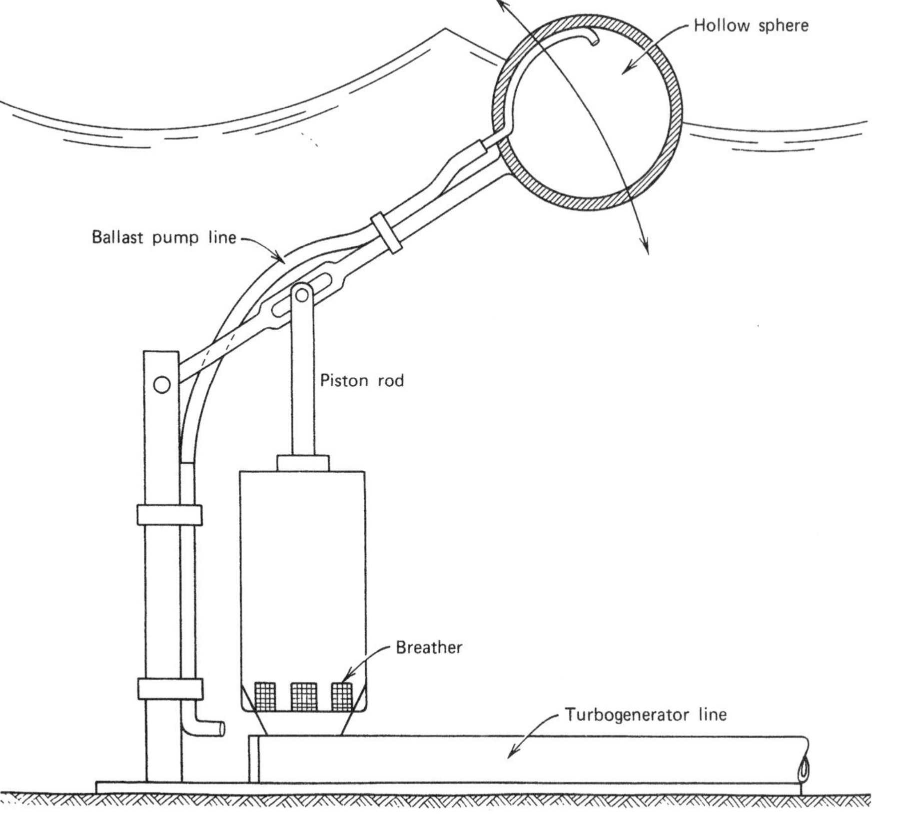

Outrigger Device (Widecrantz and Gatton)

United States Patent [19] [11] **4,172,689**

Thorsheim [45] **Oct. 30, 1979**

[54] **WAVE POWER GENERATOR**

[76] Inventor: **Ivar Thorsheim**, Kalkfjeller 15, 1370 Asker, Norway

[21] Appl. No.: **860,554**

[22] Filed: **Dec. 14, 1977**

[51] **Int. Cl.^{2}** **F94B 48/06**

[52] **U.S. Cl.** **415/7**; 415/2; 60/398; 290/42; 290/53; 417/330

[58] **Field of Search** 415/2–4, 415/7; 417/330; 60/398; 290/42, 53

[56] **References Cited**

U.S. PATENT DOCUMENTS

1,037390	9/1912	Wirt	417/330
1,081,867	12/1913	Rousseau	415/7
1,338,326	4/1920	Peck	415/7
1,476,229	12/1923	Suess	415/7
3,965,679	6/1976	Paradiso	415/2
4,036,563	7/1977	Tornkvist	60/398
4,078,871	3/1978	Perkins	60/398

FOREIGN PATENT DOCUMENTS

605673	8/1935	Fed. Rep. of Germany	415/2
2518405	11/1976	Fed. Rep. of Germany	415/2
559239	9/1923	France	415/7

Primary Examiner—Everette A. Powell, Jr.
Attorney, Agent, or Firm—Watson, Cole, Grindle & Watson

[57] **ABSTRACT**

A device for deriving power from the energy of water waves, such as ocean waves, includes a buoyant support base, such as a raft, adapted for floating in open waters and having a keel extending from its bottom surface for stabilizing same in a horizontal position. A plurality of open funnels are mounted on the top surface of the base for collecting a wave and directing it into a manifold mounted at the small ends of the funnels for operating a turbine generator in communication with the manifold. The bottom walls of the funnels extend outwardly of their larger ends so as to define artificial shoals lying beneath the bottom surface of the raft and thus beneath the surface of the water. The shoals cause the waves to break into the funnels, the broken waves tumbling along the funnels and increasing in speed upon movement toward the smaller ends thereof. The smaller ends of the funnels are staggered relative to one another so as to funnel the water into the manifold at different intervals for constant operation of the blades of the generator.

6 Claims, 5 Drawing Figures

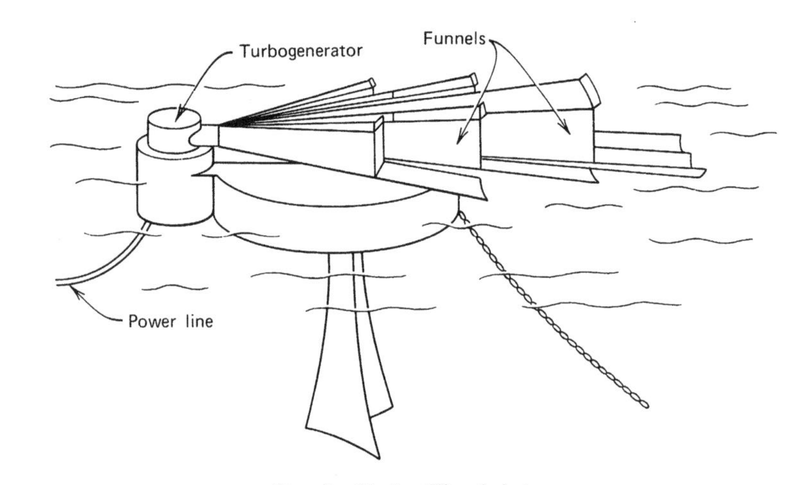

Focusing Device (Thorsheim)

United States Patent [19] [11] **4,173,517**

Salomon et al. [45] **Dec. 11, 1979**

[54] **PROCESS FOR CONVERSION OF OCEAN WAVE ENERGY INTO ELECTRIC POWER AND APPARATUS**

[75] Inventors: **Robert E. Salomon**, Dresher; **Susan M. Harding**, Bala Cynwyd, both of Pa.

[73] Assignee: **Temple University**

[21] Appl. No.: **900,668**

[22] Filed: **Apr. 27, 1978**

[51] **Int. Cl.²** **F03B 13/10**
[52] **U.S. Cl.** **290/53**; 290/42
[58] **Field of Search** 290/42, 53

[56] **References Cited**

U.S. PATENT DOCUMENTS

3,064,137 11/1962 Corbett, Jr. et al. .. 290/53
3,546,473 12/1970 Rich 290/42
3,870,893 3/1975 Mattera 290/53
3922,739 12/1975 Babintsev ..... 290/42 X

OTHER PUBLICATIONS

"A Hydrogen Electrode in Ice," by Krishnan et al. in the Journal of Physical Chemistry, vol. 70, No. 5, pp. 1595–1597, 1 May 1966.

Primary Examiner—Gene Z. Rubinson
Assistant Examiner—W. E. Duncanson, Jr.
Attorney, Agent, or Firm—Paul & Paul

[57] **ABSTRACT**

A method and apparatus is provided for direct conversion of ocean wave energy into electric power. The apparatus has no moving parts, and uses wave motion to vary the pressure of hydrogen gas in one of the cavities of a two-cavity chamber. The resulting imbalance of pressures in the cavities is relieved by conduction of hydrogen ions through a protonic conductor separating the cavities, and by conduction of electrons through an external circuit, enabling hydrogen gas to be formed on the low-pressure side of the chamber. The conduction of electrons constitutes a usable electric current. Virtually no hydrogen is consumed in this power generation process.

19 Claims, 4 Drawing Figures

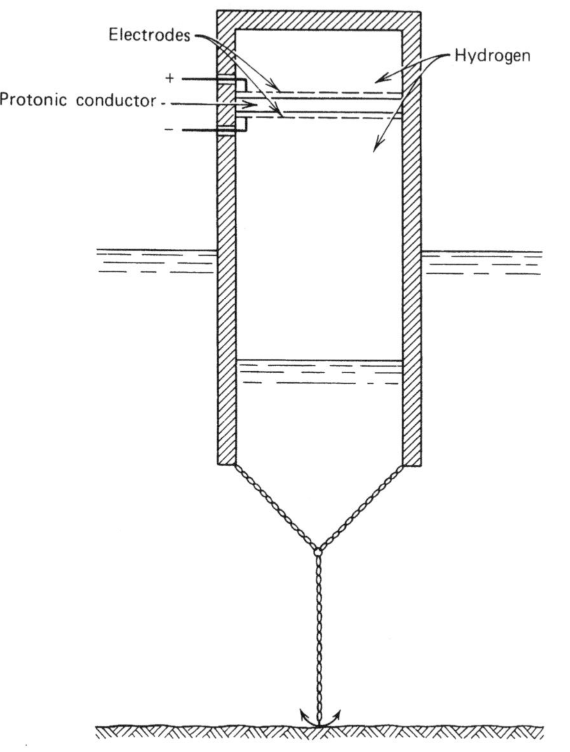

Energy Conversion (Solomon and Harding)

Appendix C

Glossary

Accretion The buildup of sand on a beach.

Added Mass The surrounding water mass excited by body motions.

Airy's Theory The theory that predicts linear or sinusoidal waves.

Antenna Effect Focusing of wave energy by constructive interference of radiant and incident waves.

Ballast Material added to a floating body to increase the weight of the body.

Breaking Wave A wave for which the water particle velocity at the crest and phase velocity are equal.

Breakwater An offshore structure designed to prematurely break waves.

Cavity Resonance A condition that occurs when the natural frequency of the water column in a cavity equals the wave frequency.

Contours Lines connecting points of constant depth on the seafloor.

Contouring Rafts Two or more rafts hinged together that follow the water surface.

Counter-rotating Turbine A turbine having runners rotating in opposite directions.

Deep Water Wave A wave for which half its length is less than the water depth.

Diffraction The distribution of wave energy in the lee of a barrier.

Dispersion The natural separation of waves according to frequency.

Double-Acting Turbine A turbine that operates in alternating flows.

Double-Peaked Spectrum A spectrum that has both a wind-wave peak and a swell peak.

Duration The time period of a wind storm.

Dynamic Positioning Position keeping by use of thrusters.

Efficiency The ratio of converted energy (power) to available energy (power).

Embedment Anchor An anchor designed to implant itself into the seafloor.

Energy-Intensive Product A manufactured device or substance requiring high energy in its production.

Erosion The loss of sand on a beach.

Fetch The length of a wind storm.

Focal Point A point at which wave power is focused.

Focusing The process of concentrating power from a broad crest width to a smaller width.

Fully Developed Sea A wind-generated sea with unchanging statistical properties.

Group Velocity The velocity of a patch of waves having the same period.

Heaving The vertical motion of a floating body.

Intermediate Water Waves Waves characterized by elliptical partical paths, the major axes of which decrease with depth.

Kaimai The barge involved in the first international wave energy conversion test in the Sea of Japan.

Kinetic Energy The energy due to particle motions in the wave.

Lens Focusing Focusing by refraction caused by a lens-shaped submerged structure.

Linear Wave A wave characterized by a sinusoidal profile.

Littoral Drift The material moved due to wave action.

Littoral Processes Those processes involving the movement of littoral drift.

Mean Water Level (MWL) The position halfway between a crest and a trough.

Monochromatic Wave A wave that has a single frequency.

Natural Frequency The characteristic frequency of a cyclic motion.

Node The intersection of the free surface of a wave and the SWL.

Nonlinear Wave A wave that has a shallow broad trough and a narrow crest.

Ocean Thermal Energy Conversion (OTEC) The conversion of the energy of vertical thermal gradients into usable energy.

One-Seventh Power Law A turbulent viscous profile equation in which the flow velocity varies as the one-seventh power of the distance from the surface.

Orthogonal or Wave Ray A line drawn tangential to successive phase velocities of a wave.

Period The time lapse between the arrivals of two successive crests of a single wave.

Phase Velocity or Celerity The velocity of a traveling wave.

Pitching The rotational motion of a floating body about an athwartship axis.

Potential Energy The energy of a particle or body due to position.

Quasi-Steady A term used to describe a phenomenon slow enough to be approximately analyzed by steady-state techniques.

Radiation Resistance The damping on a heaving, pitching, or rolling body caused by the energy removed by radiant waves created by the body motions.

Random Sea A sea composed of waves of various heights and periods that have no apparent relationships in the time domain.

Rectify To change an alternating phenomenon into a nonalternating one.

Reflected Wave A wave with a phase velocity in the direction opposite to that of the incident wave.

Refraction The process whereby a wave is turned because of a change in the phase velocity.

Resonance The condition that occurs when a natural frequency of motion is equal to the frequency of excitation.

Rolling The rotational motion of a floating body about a longitudinal axis.

Runner The moving (rotating) component of a turbine.

Sea Anchor An anchor used in deep water that relies on hydrodynamic drag for holding.

Shaddow Zone The region in the lee of a structure into which diffracted wave energy passes.

Shallow Water Wave A wave characterized by elliptical particle orbits having constant major axes.

Shoaling The condition in which the properties of the wave are affected by restricted waters.

Significant Wave Height The average height of the one-third highest waves.

Significant Wave Period The average period of the one-third highest waves.

Snell's Law The mathematical relationship describing refraction.

Solitary Wave A single shallow water wave that has an extremely low height to length ratio. The free-surface of the wave is entirely above the SWL.

Spar Buoy A long, narrow buoy that has a small waterplane area.

Standing Wave A wave that has zero-phase velocity resulting from reflection.

Station Keeping Maintaining operational position while on site.

Steepness The wave height to length ratio.

Still-Water-Level (SWL) The position of the free surface in a dead calm.

Stokes' Wave A nonlinear wave described by Stokes' expansion theory.

Superposition The combination of two or more linear waves to obtain additional wave patterns.

Surf Zone The region between the outermost breaking wave and the landward extend of the wave uprush.

Swell The long (high-period) waves resulting from far-distant storms.

Tidal Range The height of the tidal wave.

Thruster A propulsion device used for station keeping.

Tombolo A sand bar formed in the lee of a breakwater extending from the shore to the breakwater.

Traveling Wave A wave that has a nonzero-phase velocity.

Velocity Potential A mathematical artifice used to obtain the fluid velocity components in an irrotational flow.

Waterplane Area The area of a floating body at the intersection of the SWL.

Wind Waves Waves resulting from wind shear, pressure fluctuations, and turbulence.

Wave Crest The highest point of a wave.

Wave Frequency The number of waves per unit time—the inverse of the wave period.

Wave Height The vertical distance between the crest and the trough.

Wavelength The length between two successive wave crests.

Wave Setup The rise in the MWL as waves approach the shore due to mass transport by the waves.

Wave Spectrum The relationship between wave energy or power and wave frequency or period.

Wave Trough The lowest point of the surface wave.

Appendix D

Conversion Factors

Length:
1 foot (ft) = 0.3048 meter (m)
1 mile = 1.609 kilometers (km)
1 mile = 0.8684 nautical mile (nautical mile)
1 inch (in) = 0.0254 meter (m)

Surface:
1 square foot (ft^2) = 0.0929 square meter (m^2)
1 square mile ($mile^2$) = 2.59 square kilometers (km^2)
1 square inch (in^2) = 6.452×10^{-4} square meters (m^2)

Volume:
1 cubic foot (ft^3) = 0.0283 cubic meter (m^3)

Force or Weight:
1 pound (lb) = 4.448 newtons (N)
1 short ton = 0.907 metric ton
1 short ton = 2000 pounds (lb)

Mass:
1 slug = 1 pound-second^{2}per foot (lb-sec^2/ft) = 14.606 kilograms (kg)

Work or Energy:
1 foot-pound (ft-lb) = 1.356 newton-meters (N-m)
= 1.356 joules (J)

Power:
1 pound-foot per second (lb-ft/sec) = 1.356×10^{-3} kilowatt (kW)
1 horsepower (hp) = 0.7457 kilowatt (kW)
1 watt (W) = 1 joule per second (J/sec)

Pressure:
1 pound per square inch = 144 pounds per square foot
1 pound per square foot = 47.88 newtons per square meter (N/m^2)

Index

Added-mass, 47-49
Added-mass moment of inertia, 48-51, 181, 182
Aesthetic consideration, 170
Airy wave theory, 7
Anchoring:
- Boss, 192
- breakout, 192
- Danforth, 192
- deadweight, 192
- drag embedment, 192
- embedment, 181, 182
- holding power, 192, 193
- Navy stockless, 192
- pile, 192
- plate, 192
- Stato, 192, 193
- Stimson, 192

Ballasting, 70, 124
Battery, floating, 164, 166
Bernoulli's equation, 71, 147
Breaking wave, 13, 17, 20, 78, 126
Breakwater, 175, 176
Bretschneider spectrum, 23, 172

Cavity resonance, 61, 63, 67, 143
- design condition, 66
- optimum, 70

Coastal processes:
- accretion, 175
- beach stability, 175
- erosion, 175
- littoral drift, 175
- longshore power, 178-180
- tombolo formation, 175, 176
- sand bar, 175

Coastal zone operation, 170, 175-180, 190-194
Cockerall/Hagen Raft, 101-110, 202-225
Continuity equation, 62, 147
Convection velocity, 16

DAM-ATOLL, 124-129
- fluid flywheel, 126
- radial diffuser, 126

Deep water wave approximation, 9, 14, 28, 75, 85, 92
Diffraction, 29, 36-44
- coefficient, 35, 37, 39, 40-43

Dispersion, 47

Economic considerations, 4
Electricity:
- a.c. cable, 164
- cable joints, 164
- direct current (d.c.), 137, 140
  - cable, 163, 164-166
- instantaneous current, 143
- instantaneous power, 144, 150, 151, 152
- instantaneous voltage, 142
- power grid, 164
- transmission loss, 164
- transmitted power, 171

Electrochemical bridge, 166
Energy conversion, 202-225
Energy-intensive products, 137, 166-168
Energy storage, 148, 164, 166
Energy transfer, 163-167
Environmental considerations, 170-180

Fetch, 171, 172
Float:
- circular, 57
- rectangular, 57

Focusing (wave), 60, 117-133
- antenna, 117-124, 175, 184
- island, 117, 124
- lens, 117, 124-133

Force (heaving), 47
Free-surface displacement, 7, 8, 32, 33
Frequency:
- heaving, 47, 181, 182, 183
- pitching, 50, 182, 183

Froude numbers, 161

Fully-developed sea, 23, 171

Generator, mechanical-drive, 137
Group velocity, 14, 120

Heaving, 46-60
  efficiency, 59
  energy, 55
  device, 196-201, 202-225
Hydraulic equation, 76
Hydrazine, 166

Intermediate water waves, 12
International Energy Agency (IEA), 70, 71

Kaimei, 71, 143, 182

Lens maker's equation, 130
Linear inductance generator, 147, 149-152
Linear wave theory, 7, 16, 22, 71
Liquified ammonia, 166
Liquified hydrogen, 166
Lithium battery, 166, 167

Magnification factor, 54, 55, 56
Mass density of salt water, 13
Mean Water Level (MWL), 15
Moment (pitching), 47
Momentum equation, 147
Moorings, 180-194
  catenary, 190-190
  effects, 96-98
  materials, 185-186
    metals, 185
    natural nonmetals, 185
    synthetics, 185
  multi-component, 186
  rigid, 190
  single point, 182
  slack, 181, 186, 190, 191
  spring, constant, 183
  tension, 181, 183, 186

Ocean Thermal Energy Conversion (OTEC),
    3, 164, 166, 167
One-Seventh Power Law, 173
Open ocean operation, 170-175, 181-190
Orthogonal, 125, 129
  logarithmic spiral, 125
Oscillating water column, 71, 72, 202-225
Outrigger device, 196-201, 202-225

Particle Motion Device, 196-201, 202-225
Phase velocity (celerity), 9, 10, 20

Pierson-Moskowitz spectrum, 23, 25
Piezoelectricity, 153-158
  compressional materials, 153
  crystals, 153
  design wave, 153
  dielectric permittivity, 154
  efficiency, 158
  Honeywell Composite, 155
  Kureha Piezofilm, 155
  stack, 153, 154
  strain constant, 153
  stress constant, 153
  transverse materials, 153

Pitching, 46-60
  device, 196-201, 202-225
  energy, 56, 57
  power, 58
Pneumatic Device, 196-201, 202-225
Point absorber, 175
Pressure, 21
  device, 196-201, 202-225
  dynamic, 71, 72
  hydrostatic, 71
Protonic conduction, 158-160
  electrochemical hydrogen concentration
    cell, 158
  Faraday constant, 159
  high pressure cell, 160
  Nafion, 159
  Nerust equation, 158

Radiation resistance, 120
Reflection, 28, 32-36
  oblique, 36
Refraction, 28-32
  coefficient, 29, 30, 31, 32
Resonance, 47, 51, 57, 139, 175
  harbor, 114-116
Riser, 164, 185
Russel rectifier, 110-117

Salinity gradient energy, 3
Salter's Nodding Duck, 90-101, 196-201,
    202-225
  design condition, 92, 98
  design wave, 92
  inertia, 94
  natural frequency, 91, 92, 94, 96
  stiffness, 93
Satellite wave energy converters, 184
Scale factor, 160-161
Shadow zone, 36
Shallow water approximation, 78, 84, 155

Shoaling, 4, 10, 29
  coefficient, 29, 32
Significant wave, 7, 25
  height, 25, 172, 173
  period, 172, 173
Snell's law, 28, 29
Solar energy, 1
Solitary wave, 16-19, 81
Still Water Level (SWL), 12, 15, 20, 84
Stokes' theory, 16, 19-21
Strouhal number, 161
Superposition, 33
Surf zone, 20, 79
Surging device, 79-84, 196-201, 202-225
  compliant flap, 86-90
  power, 82
Swell, 7, 47, 72, 107
Synthetic fuels, 166

Thruster, 184-185
Tidal energy, 3
Turbines:
  blade coefficient, 148
  double-acting, 63
  McCormick's, 144
  power, 147
  rectifying flaps, 143
  Wells', 143

Uniform pressure wave pump, 72
  closed system, 73, 74
  constant force, 76-77
  efficiency, 75, 77
  open system, 73, 74
  work, 75, 78

Velocity potential, 71, 147

Water particle velocity, 11, 16, 20
Water wheel, 85, 86
Wildlife breeding, 170
Wind-Mill battery system, 184, 185
Wind velocity, 23, 24, 25, 26, 172, 173
Wave:
  crest, 13
  energy, 1, 3, 13, 18, 20, 24, 33, 34
  frequency, 8
  height, 7, 35
  kinematic properties, 9
  left-running, 32
  length, 7, 8
  linear, 7, 16, 22, 46, 71
  monochromatic, 22
  node, 13
  nonlinear, 15-21
  number, 9, 72, 106
  period, 8, 24, 173
  power, 13, 18, 83, 93, 97, 119
  random, 22-26, 47, 99-101
  rider, 53
  right-running, 32
  set-up, 19
  spectrum, 22
  standing, 33, 144